NUMERICAL HEAT TRANSFER AND FLUID FLOW

Suhas V. Patankar

Professor of Mechanical Engineering
University of Minnesota

Taylor & Francis

Publishers since 1798

NUMERICAL HEAT TRANSFER AND FLUID FLOW

Library of Congress Cataloging in Publication Data

Patankar, S V
    Numerical heat transfer and fluid flow.

    (Series in computational methods in mechanics and
thermal sciences)
    Bibliography: p.
    Includes index.
    1. Heat—Transmission. 2. Fluid dynamics.
3. Numerical analysis. I. Title. II. Series.
QC320.P37   536'.2   79-28286
ISBN 0-89116-522-3

This book was set in Press Roman by Hemisphere Publishing Corporation.
The editors were Mary A. Phillips and Edward M. Millman; the production
supervisor was Rebekah McKinney; and the typesetter was Sandra F. Watts.
BookCrafters, Inc., was printer and binder.

U H University of
Hertfordshire

College Lane, Hatfield, Herts AL10 9AB

Learning and Information Services

For renewal of Standard and One Week Loans,
please visit the website: **http://www.voyager.herts.ac.uk**

This item must be returned or the loan renewed by the due date.
The University reserves the right to recall items from loan at any time.
A fine will be charged for the late return of items.

NUMERICAL
HEAT TRANSFER
AND FLUID FLOW

Series in Computational and Physical Processes in Mechanics and Thermal Sciences

(*Formerly the Series in Computational Methods in Mechanics and Thermal Sciences*)

W. J. Minkowycz and E. M. Sparrow, *Editors*

Anderson, Tannehill, and Pletcher, Computational Fluid Mechanics and Heat Transfer
Aziz and Na, Perturbation Methods in Heat Transfer
Baker, Finite Element Computational Fluid Mechanics
Beck, Cole, Haji-Shiekh, and Litkouhi, Heat Conduction Using Green's Functions
Chung, Editor, Numerical Modeling in Combustion
Jaluria and Torrance, Computational Heat Transfer
Patankar, Numerical Heat Transfer and Fluid Flow
Pepper and Heinrich, The Finite Element Method: Basic Concepts and Applications
Shih, Numerical Heat Transfer

PROCEEDINGS

Chung, Editor, Finite Elements in Fluids: Volume 8
Haji-Sheikh, Editor, Integral Methods in Science and Engineering-90
Shih, Editor, Numerical Properties and Methodologies in Heat Transfer: Proceedings of the Second National Symposium

FORTHCOMING TITLES

Pepper and Heinrich, The Finite Element Method: Advanced Concepts
Pletcher, Anderson, and Tannehill, Computational Fluid Mechanics and Heat Transfer, Second Edition
Willmott, Modeling and Dynamics of Regenerative Heat Transfer

To my wife Rajani

CONTENTS

3 DISCRETIZATION METHODS   25

4 HEAT CONDUCTION   41

5 CONVECTION AND DIFFUSION   79

6   CALCULATION OF THE FLOW FIELD   113

7   FINISHING TOUCHES   139

PREFACE

In 1972, I taught an informal course on numerical solution of heat transfer and fluid flow to a small group of research workers at Imperial College, London. Later, the material was expanded and formalized for presentation in graduate courses at the University of Waterloo in Canada (in 1974), at the Norwegian Institute of Technology, Trondheim (in 1977), and at the University of Minnesota (in 1975, 1977, and 1979). During the last two years, I have also presented the same material in a short-course format at ASME national meetings. The enthusiastic response accorded to these courses has encouraged me to write this book, which can be used as a text for a graduate course as well as a reference book for computational work in heat transfer and fluid flow.

Although there is an extensive literature on computational thermofluid analysis, the newcomer to the field has insufficient help available. The graduate student, the researcher, and the practicing engineer must struggle through journal articles or be content with elementary presentation in books on numerical analysis. Often, it is the subtle details that determine the success or failure of a computational activity; yet, the practices that are learned through experience by successful computing groups rarely appear in print. A consequence is that many workers either give up the computational approach after many months of frustrating pursuit or struggle through to the end with inefficient computer programs.

Being aware of this situation, I have tried to present in this book a self-contained, simple, and practical treatment of the subject. The book is introductory in style and is intended for the potential practitioner of numerical heat transfer and fluid flow; it is not designed for the experts in the subject area. In developing the numerical techniques, I emphasize physical significance rather than mathematical manipulation. Indeed, most of the mathematics used here is limited to simple algebra. The result is that, whereas the book enables the reader to travel all the way to the present-day frontier of the subject, the journey takes place through delightfully simple and illuminating physical concepts and considerations. In teaching the material with such an approach, I have often been pleasantly surprised by the fact that the students not only learn about numerical methods but also develop a better appreciation of the relevant physical processes.

As a user of numerical techniques, I have come to prefer a certain family of methods and a certain set of practices. This repertoire has been collected partly from the literature and subsequently has been enriched, adapted, and modified. Thus, since a considerable amount of sorting and sifting of available methods has already taken place (albeit with my own bias), I have limited the scope of this book to the set of methods that I wish to recommend. I do not attempt to present a comparative study of all available methods; other methods are only occasionally mentioned when they serve to illuminate a specific feature under consideration. In this sense, this book represents my personal view of the subject. Although I am, of course, enthusiastic about this viewpoint, I recognize that my choices have been influenced by my background, personal preferences, and technical objectives. Others operating in different environments may well come to prefer alternative approaches.

To illustrate the application of the material, problems are given at the end of some chapters. Most of the problems can be solved by using a pocket calculator, although some of them should be programmed for a digital computer. The problems are not meant for testing the student reader, but are included primarily for extending and enriching the learning process. They suggest alternative techniques and present additional material. At times, in my attempt to give a hint for the problem solution, I almost disclose the full answer. In such cases, arriving at the correct answer is not the main objective; the reader should focus on the message that the problem is designed to convey.

This book carries the description of the numerical method to a point where the reader could begin to write a computer program. Indeed, the reader should be able to construct computer programs that generate the kind of results presented in the final chapter of the book. A range of computer programs of varying generality can be designed depending upon the nature of the problems to be solved. Many readers might have found it helpful if a representative computer program were included in this book. I did consider the possibility. However, the task of providing a reasonably general computer program, its detailed description, and several examples of its use seemed so formidable that it would have considerably delayed the publication of this book. For the time being, I have included a section on the preparation and testing of a computer program (Section 7.4), where many useful procedures and practices gathered through experience are described.

The completion of this book fulfills a desire and a dream that I have held for a number of years. It was in 1971 that Professor D. Brian Spalding and I planned a book of this kind and wrote a preliminary outline for it. Further progress, however, became difficult because of the geographical distance between us and because of our involvement in a variety of demanding activities. Finally, a joint book seemed impracticable, and I proceeded to convert my lecture notes into this textbook. The present book has some resemblance to the joint book that we had planned, since I have made liberal use of Spalding's lectures and writings. His direct involvement, however, would have made this book much better.

In this undertaking, I owe the greatest debt to Professor Spalding. He introduced me to the fascinating world of computational methods. The work that we accomplished together represents the most delightful and creative experience of my professional life. The influence of his ideas on my thinking can be seen throughout this book. The concepts of "one-way" and "two-way" coordinates (and the terms themselves) are the product of his imagination. It was he who organized all the relevant physical processes through a general differential equation of a standard form. Above all, our rapid progress in computational work has resulted from Spalding's vision and conviction that one day all practical situations will become amenable to computer analysis.

I wish to record my sincere thanks to Professor D. Brian Spalding for his creative influence on my professional activities, for continued warm friendship, and for his direct and indirect contributions to this book.

Professor Ephraim M. Sparrow has been my most enthusiastic supporter in the activity of writing this book. His interest began even earlier when he attended my graduate course on the subject. I have greatly benefited from his questions and subsequent discussions. He spent countless hours in reading the manuscript of this book and in suggesting changes and improvements. It is due to his critical review that I have been able to achieve some measure of clarity and completeness in this book. I am very grateful to him for his active interest in this work and for his personal interest in me.

A number of other colleagues and friends have also provided constant inspiration through their special interest in my work. In particular, I wish to thank Professor Richard J. Goldstein for his support and encouragement and Professor George D. Raithby for many stimulating discussions. My thanks are also due to the many students in my graduate courses, who have contributed significantly to this book through their questions and discussions and through their enthusiasm and response. I am grateful to Mrs. Lucille R. Laing, who typed the manuscript so carefully and cheerfully. I would like to thank Mr. William Begell, President of Hemisphere Publishing Corporation, for his personal interest in publishing this book and the staff at Hemisphere for their competent handling of this project.

My family has been very understanding and supportive during my writing activity; now that the writing is over, I plan to spend more time with my wife and children.

NUMERICAL
HEAT TRANSFER
AND FLUID FLOW

INTRODUCTION

1.1 SCOPE OF THE BOOK

Importance of heat transfer and fluid flow. This book is concerned with heat and mass transfer, fluid flow, chemical reaction, and other related processes that occur in engineering equipment, in the natural environment, and in living organisms. That these processes play a vital role can be observed in a great variety of practical situations. Nearly all methods of power production involve fluid flow and heat transfer as essential processes. The same processes govern the heating and air conditioning of buildings. Major segments of the chemical and metallurgical industries use components such as furnaces, heat exchangers, condensers, and reactors, where thermofluid processes are at work. Aircraft and rockets owe their functioning to fluid flow, heat transfer, and chemical reaction. In the design of electrical machinery and electronic circuits, heat transfer is often the limiting factor. The pollution of the natural environment is largely caused by heat and mass transfer, and so are storms, floods, and fires. In the face of changing weather conditions, the human body resorts to heat and mass transfer for its temperature control. The processes of heat transfer and fluid flow seem to pervade all aspects of our life.

Need for understanding and prediction. Since the processes under consideration have such an overwhelming impact on human life, we should be able to deal with them effectively. This ability can result from an understanding of the nature of the processes and from methodology with which to predict them quantitatively. Armed with this expertise, the designer of an engineering device can ensure the desired performance—the designer is able to choose the

1

optimum design from among a number of alternative possibilities. The power of prediction enables us to operate existing equipment more safely and efficiently. Predictions of the relevant processes help us in forecasting, and even controlling, potential dangers such as floods, tides, and fires. In all these cases, predictions offer economic benefits and contribute to human well-being.

Nature of prediction. The prediction of behavior in a given physical situation consists of the values of the relevant variables governing the processes of interest. Let us consider a particular example. In a combustion chamber of a certain description, a complete prediction should give us the values of velocity, pressure, temperature, concentrations of the relevant chemical species, etc., throughout the domain of interest; it should also provide the shear stresses, heat fluxes, and mass flow rates at the confining walls of the combustion chamber. The prediction should state how any of these quantities would change in response to proposed changes in geometry, flow rates, fluid properties, etc.

Purpose of the book. This book is primarily aimed at developing a general method of prediction for heat and mass transfer, fluid flow, and related processes. As we shall shortly see, among the different methods of prediction, the *numerical* solution offers great promise. In this book, we shall construct a numerical method for predicting the processes of interest.

As far as possible, our aim will be to design a numerical method having complete generality. We shall, therefore, refrain from accepting any final restrictions such as two-dimensionality, boundary-layer approximations, and constant-density flow. If any restrictions are temporarily adopted, it will be for ease of presentation and understanding and not because of any intrinsic limitation. We shall begin the subject at a very elementary level and, from there, travel nearly to the frontier of the subject.

This ambitious task cannot, of course, be accomplished in a modest-sized book without leaving out a number of important topics. Therefore, the mathematical formulation of the equations that govern the processes of interest will be discussed only briefly in this book. For the complete derivation of the required equations, the reader must turn to standard textbooks on the subject. The mathematical models for complex processes like turbulence, combustion, and radiation will be assumed to be known or available to the reader. Even in the subject of numerical solution, we shall not survey all available methods and discuss their merits and demerits. Rather, we shall focus attention on a particular family of methods that the author has used, developed, or contributed to. Reference to other methods will be made only when this serves to highlight a certain issue. While a general formulation will be attempted, no special attention will be given to supersonic flows, free-surface flows, or two-phase flows.

An important characteristic of the numerical methods to be developed in this book is that they are strongly based on physical considerations, not just on mathematical manipulations. Indeed, nothing more sophisticated than simple algebra and elementary calculus is used. A significant advantage of this strategy is that the reader, while learning about the numerical methods, develops a deeper understanding of, and insight into, the underlying physical processes. This appreciation for physical significance is very helpful in analyzing and interpreting computed results. But, even if the reader never performs numerical computations, this study of the numerical methods will provide—it is interesting to note—a greater feel for the physical aspects of heat transfer and fluid flow. Further, the physical approach will equip the reader with general criteria with which to judge other existing and future numerical methods.

1.2 METHODS OF PREDICTION

Prediction of heat transfer and fluid-flow processes can be obtained by two main methods: experimental investigation and theoretical calculation. We shall briefly consider each and then compare the two.

1.2-1 Experimental Investigation

The most reliable information about a physical process is often given by actual measurement. An experimental investigation involving full-scale equipment can be used to predict how identical copies of the equipment would perform under the same conditions. Such full-scale tests are, in most cases, prohibitively expensive and often impossible. The alternative then is to perform experiments on small-scale models. The resulting information, however, must be extrapolated to full scale, and general rules for doing this are often unavailable. Further, the small-scale models do not always simulate all the features of the full-scale equipment; frequently, important features such as combustion or boiling are omitted from the model tests. This further reduces the usefulness of the test results. Finally, it must be remembered that there are serious difficulties of measurement in many situations, and that the measuring instruments are not free from errors.

1.2-2 Theoretical Calculation

A theoretical prediction works out the consequences of a mathematical model, rather than those of an actual physical model. For the physical processes of interest here, the mathematical model mainly consists of a set of differential equations. If the methods of classical mathematics were to be used for solving these equations, there would be little hope of predicting many phenomena of practical interest. A look at a classical text on heat conduction or fluid mechanics leads to the conclusion that only a tiny fraction of the range of practical problems can be solved in closed form. Further, these solutions often

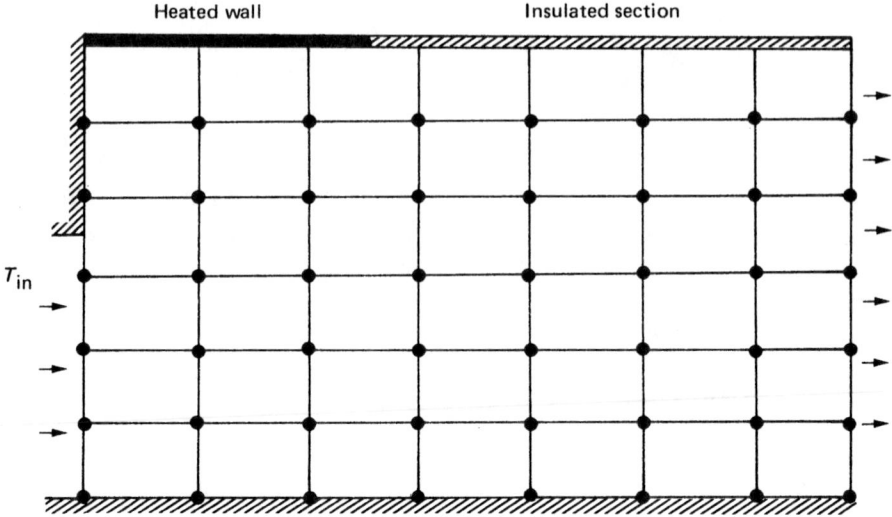

Figure 1.1 Grid layout for a numerical solution for the temperature field.

contain infinite series, special functions, transcendental equations for eigen-values, etc., so that their numerical evaluation may present a formidable task.[*]

Fortunately, the development of numerical methods and the availability of large digital computers hold the *promise* that the implications of a mathematical model can be worked out for almost any practical problem. A preliminary idea of the numerical approach to problem solving can be obtained by reference to Fig. 1.1. Suppose that we wish to obtain the temperature field in the domain shown. It may be sufficient to know the *values* of temperature at discrete points of the domain. One possible method is to imagine a grid that fills the domain, and to seek the values of temperature at the grid points. We then construct and solve *algebraic* equations for these unknown temperatures. The simplification inherent in the use of algebraic equations rather than differential equations is what makes numerical methods so powerful and widely applicable.

1.2-3 Advantages of a Theoretical Calculation

We shall now list the advantages that a theoretical calculation offers over a corresponding experimental investigation.

[*]It is not implied here that exact analytical solutions are without practical value. Indeed, as we shall see later, some features of numerical methods are constructed by the use of simple analytical solutions. Further, there is no better way of checking the accuracy of a numerical method than by comparison with an exact analytical solution. However, there seems to be little doubt that the methods of classical mathematics do not offer a practical way of solving complex engineering problems.

Low cost. The most important advantage of a computational prediction is its low cost. In most applications, the cost of a computer run is many orders of magnitude lower than the cost of a corresponding experimental investigation. This factor assumes increasing importance as the physical situation to be studied becomes larger and more complicated. Further, whereas the prices of most items are increasing, computing costs are likely to be even lower in the future.

Speed. A computational investigation can be performed with remarkable speed. A designer can study the implications of hundreds of different configurations in less than a day and choose the optimum design. On the other hand, a corresponding experimental investigation, it is easy to imagine, would take a very long time.

Complete information. A computer solution of a problem gives detailed and complete information. It can provide the values of *all* the relevant variables (such as velocity, pressure, temperature, concentration, turbulence intensity) *throughout* the domain of interest. Unlike the situation in an experiment, there are few inaccessible locations in a computation, and there is no counterpart to the flow disturbance caused by the probes. Obviously, no experimental study can be expected to measure the distributions of all variables over the entire domain. For this reason, even when an experiment is performed, there is great value in obtaining a companion computer solution to supplement the experimental information.

Ability to simulate realistic conditions. In a theoretical calculation, realistic conditions can be easily simulated. There is no need to resort to small-scale or cold-flow models. For a computer program, there is little difficulty in having very large or very small dimensions, in treating very low or very high temperatures, in handling toxic or flammable substances, or in following very fast or very slow processes.

Ability to simulate ideal conditions. A prediction method is sometimes used to study a basic phenomenon, rather than a complex engineering application. In the study of a phenomenon, one wants to focus attention on a few essential parameters and eliminate all irrelevant features. Thus, many idealizations are desirable—for example, two-dimensionality, constant density, an adiabatic surface, or infinite reaction rate. In a computation, such conditions can be easily and exactly set up. On the other hand, even a very careful experiment can barely approximate the idealization.

1.2-4 Disadvantages of a Theoretical Calculation

The foregoing advantages are sufficiently impressive to stimulate enthusiasm about computer analysis. A blind enthusiasm for any cause is, however, undesirable. It is useful to be aware of the drawbacks and limitations.

As mentioned earlier, a computer analysis works out the implications of a mathematical model. The experimental investigation, by contrast, observes the reality itself. The validity of the mathematical model, therefore, limits the usefulness of a computation. In this book, we shall be concerned only with computational methods and not with mathematical models. Yet, we must note that the *user* of the computer analysis receives an end product that depends on both the mathematical model and the numerical method. A perfectly satisfactory numerical technique can produce worthless results if an inadequate mathematical model is employed.

For the purpose of discussing the disadvantages of a theoretical calculation, it is, therefore, useful to divide all practical problems into two groups:

Group A: Problems for which an adequate mathematical description can be written. (Examples: heat conduction, laminar flows, simple turbulent boundary layers.)
Group B: Problems for which an adequate mathematical description has not yet been worked out. (Examples: complex turbulent flows, certain non-Newtonian flows, formation of nitric oxides in turbulent combustion, some two-phase flows.)

Of course, the group into which a given problem falls will be determined by what we are prepared to consider as an "adequate" description.

Disadvantages for Group A. It may be stated that, for most problems of Group A, the theoretical calculation suffers from no disadvantages. The computer solution then represents an alternative that is highly superior to an experimental study. Occasionally, however, one encounters some disadvantages. If the prediction has a very limited objective (such as finding the overall pressure drop for a complicated apparatus), the computation may not be less expensive than an experiment. For difficult problems involving complex geometry, strong nonlinearities, sensitive fluid-property variations, etc., a numerical solution may be hard to obtain and would be excessively expensive if at all possible. Extremely fast and small-scale phenomena such as turbulence, if they are to be computed in all their time-dependent detail by solving the unsteady Navier-Stokes equations, are still beyond the practical reach of computational methods. Finally, when the mathematical problem occasionally admits more than one solution, it is not easy to determine whether the computed solution corresponds to reality.

Research in computational methods is aimed at making them more reliable, accurate, and efficient. The disadvantages mentioned here will diminish as this research progresses.

Disadvantages for Group B. The problems of Group B share all the disadvantages of Group A; in addition, there is the uncertainty about the extent to which the computed results would agree with reality. In such cases, some experimental backup is highly desirable.

Research in mathematical models causes a transfer of problems from Group B into Group A. This research consists of proposing a model, working out its implications by computer analysis, and comparing the results with experimental data. Thus, computational methods play a key role in this research. A striking example of this role can be found in the recent development of turbulence models. The currently popular and widely used two-equation turbulence models are primarily based on the work of Kolmogorov (1942) and Prandtl (1945). It was, however, only in the 1970s, when computers and computational methods became more powerful, that the turbulence models were put to practical use.

1.2-5 Choice of Prediction Method

This discussion about the relative merits of computer analysis and experimental investigation is not aimed at recommending computation to the exclusion of experiment. An appreciation of the strengths and weaknesses of both approaches is essential to the proper choice of the appropriate technique.

There is no doubt that experiment is the only method for investigating a new basic phenomenon. In this sense, experiment leads and computation follows. It is in the synthesis of a number of interacting known phenomena that the computation performs more efficiently. Even then, sufficient validation of the computed results by comparison with experimental data is required. On the other hand, for the design of experimental apparatus, preliminary computations are often helpful, and the *amount* of experimentation can usually be significantly reduced if the investigation is supplemented by computation.

An optimal prediction effort should thus be a judicious combination of computation and experiment. The proportions of the two ingredients would depend on the nature of the problem, on the objectives of the prediction, and on the economic and other constraints of the situation.

1.3 OUTLINE OF THE BOOK

This book is designed to unfold the subject in a certain sequence, and the reader is urged to follow the same sequence. It will not be profitable to jump to a later chapter, as all chapters build upon the material covered in the previous ones. The problems at the end of some chapters are intended to give the reader both direct experience with and deeper understanding of the principles developed in the book.

The nine chapters that comprise this book can be grouped into three different parts of three chapters each. The first three chapters constitute the preparatory phase. Here, a preliminary discussion about the mathematical and numerical aspects is included, and the particular philosophy of the book is outlined. Chapters 4-6 contain the main development of the numerical method. The last three chapters are devoted to elucidations and applications.

Before we begin the task of numerical solution, the physical phenomena must be described via appropriate differential equations. This is outlined and discussed in Chapter 2. Of special importance in that chapter is the examination of the parabolic or elliptic nature of these equations from a physically meaningful viewpoint.

The concept of numerical solution is developed in Chapter 3, where the common procedures of constructing numerical methods are described. Among these, the method that lends itself to easy physical interpretation is chosen and illustrated by means of a very simple example. This introductory material is used to formulate general criteria in the form of four basic rules. These rules form the guideposts for the development of the numerical method in the rest of this book. Although the rules are formulated from physical considerations alone, they often lead to results that—it is interesting to observe—are normally derived from purely mathematical analysis. Furthermore, these rules guide us to better formulations that may not have been suggested by standard mathematical methods.

The construction of the numerical method begins in Chapter 4. It is carried out in three stages. Heat conduction (i.e., the general problem without the convection term) is treated in Chapter 4. Chapter 5 concentrates on the interaction of convection and conduction, with the flow field regarded as given. Finally, the calculation of the velocity field itself is dealt with in Chapter 6.

Readers who are interested in fluid flow alone, and not in heat transfer, should note that Chapter 6 is not a self-contained chapter. It describes only the *additional* features required for the fluid-flow calculation, the other details having already been given in Chapters 4 and 5. Thus, Chapter 4 does not merely deal with heat conduction; it completes much of the groundwork needed for fluid flow. The treatment of convection in Chapter 5 is also equally applicable to fluid-flow calculation. This approach—handling fluid flow through heat transfer—may be unfamiliar to some readers, but it appears to be an effective pedagogical technique. The early focus on heat transfer enables us to conduct all the preliminary discussion in terms of temperature, which is an easy-to-understand scalar variable. It also reinforces the conceptual unity between variables such as temperature and momentum, which is useful in understanding and interpreting results.

Another technique that will be in evidence in these chapters is the use of *one-dimensional* situations to construct the basic algorithm, which is then quickly generalized to multidimensional cases. The one-dimensional problem serves to keep the algebraic complication to a minimum and to focus attention on the significant issues.

Chapter 7 is a compilation of a number of elucidating remarks and suggestions that can be properly appreciated after the reader has had an overview of the method through familiarity with the first six chapters. Chapter 8 deals with calculation procedures that can be considered as special cases of the general method developed in the book. The control-volume-based finite-element method, which is briefly described in Section 8.4, is, however, an extension rather than a special case of the general method.

The last chapter serves to give the reader a taste of possible applications of the method. It contains a brief description of some of the problems solved by the author and his co-workers. This is, of course, only a very small fraction of the totality of interesting problems that are within the reach of the method. The possibilities are limited only by the imagination of the user.

TWO

MATHEMATICAL DESCRIPTION OF PHYSICAL PHENOMENA

The numerical solution of heat transfer, fluid flow, and other related processes can begin when the laws governing these processes have been expressed in mathematical form, generally in terms of differential equations. For a detailed and complete derivation of these equations, the reader should turn to a standard textbook. Our purpose here is to develop familiarity with the form and the meaning of these equations. It will be shown that all the equations of relevance here possess a common form, the identification of which is the first step toward constructing a general solution procedure. We shall also discuss some characteristics of the independent variables used in these equations.

2.1 GOVERNING DIFFERENTIAL EQUATIONS

2.1-1 Meaning of a Differential Equation

The individual differential equations that we shall encounter express a certain conservation principle. Each equation employs a certain physical quantity as its dependent variable and implies that there must be a balance among the various factors that influence the variable. The dependent variables of these differential equations are usually *specific* properties, i.e., quantities expressed on a unit-mass basis. Examples[*] are mass fraction, velocity (i.e., momentum per unit mass), and specific enthalpy.

---

[*]Temperature, which is quite frequently used as a dependent variable, is not a specific property; it arises from more basic equations employing specific internal energy or specific enthalpy as the dependent variable.

The terms in a differential equation of this type denote influences on a *unit-volume* basis. An example will make this clear. Suppose $\mathbf{J}$ denotes a flux influencing a typical dependent variable ϕ. Let us consider the control volume of dimensions dx, dy, dz shown in Fig. 2.1. The flux J_x (which is the x-direction component of $\mathbf{J}$) is shown entering one face of area $dy\,dz$, while the flux leaving the opposite face is shown as $J_x + (\partial J_x/\partial x)\,dx$. Thus, the *net* efflux is $(\partial J_x/\partial x)\,dx\,dy\,dz$ over the area of the face. Considering the contributions of the y and z directions as well and noting that $dx\,dy\,dz$ is the volume of the region considered, we have

$$\text{Net efflux per unit volume} = \frac{\partial J_x}{\partial x} + \frac{\partial J_y}{\partial y} + \frac{\partial J_z}{\partial z}$$

$$= \text{div } \mathbf{J}. \tag{2.1}$$

This interpretation of div $\mathbf{J}$ will be particularly useful to us because, as we shall see later, our numerical method will be constructed by performing a balance over a control volume.

Another example of a term expressed on a unit-volume basis is the rate-of-change term $\partial(\rho\phi)/\partial t$. If ϕ is a specific property and ρ is the density, then $\rho\phi$ denotes the amount of the corresponding *extensive* property contained in a unit volume. Thus, $\partial(\rho\phi)/\partial t$ is the rate of change of the relevant property per unit volume.

A differential equation is a compilation of such terms, each representing an influence on a unit-volume basis, and all the terms together implying a balance or conservation. We shall now take as examples a few standard differential equations, to find a general form.

2.1-2 Conservation of a Chemical Species

Let m_l denote the mass fraction* of a chemical species. In the presence of a velocity field $\mathbf{u}$, the conservation of m_l is expressed as

$$\frac{\partial}{\partial t}(\rho m_l) + \text{div }(\rho \mathbf{u} m_l + \mathbf{J}_l) = R_l. \tag{2.2}$$

Here $\partial(\rho m_l)/\partial t$ denotes, as explained earlier, the rate of change of the mass of the chemical species per unit volume. The quantity $\rho \mathbf{u} m_l$ is the convection flux of the species, i.e., the flux carried by the general flow field $\rho \mathbf{u}$. The symbol $\mathbf{J}_l$ stands for the diffusion flux, which is normally caused by the

---

*The mass fraction m_l of a chemical species l is defined as the ratio of the mass of the species l (contained in a given volume) to the total mass of the mixture (contained in the same volume).

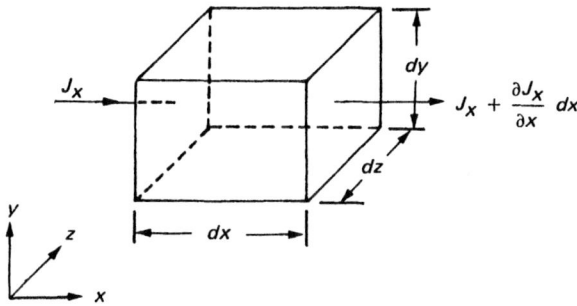

Figure 2.1 Flux balance over a control volume.

gradients of m_l. The divergence of the two fluxes (convection and diffusion) forms the second term of the differential equation. The quantity R_l on the right-hand side is the rate of generation of the chemical species per unit volume. The generation is caused by chemical reaction. Of course, R_l can have a positive or negative value depending on whether the reaction actually produces or destroys the chemical species, and R_l is zero for a nonreacting species.

If the diffusion flux J_l is expressed by the use of Fick's law of diffusion, we can write

$$J_l = -\Gamma_l \text{ grad } m_l ,\qquad (2.3)$$

where Γ_l is the diffusion coefficient. The substitution of Eq. (2.3) into (2.2) leads to

$$\frac{\partial}{\partial t} (\rho m_l) + \text{div} (\rho \mathbf{u} m_l) = \text{div} (\Gamma_l \text{ grad } m_l) + R_l .\qquad (2.4)$$

2.1-3 The Energy Equation

The energy equation in its most general form contains a large number of influences. Since we are primarily interested in the form rather than in the details of the equation, it will be sufficient to consider some restricted cases.

For a steady low-velocity flow with negligible viscous dissipation, the energy equation can be written as

$$\text{div} (\rho \mathbf{u} h) = \text{div} (k \text{ grad } T) + S_h ,\qquad (2.5)$$

where h is the specific enthalpy, k is the thermal conductivity, T is the temperature, and S_h is the volumetric rate of heat generation. The term div $(k$ grad $T)$ represents the influence of conduction heat transfer within the fluid, according to the Fourier law of conduction.

For ideal gases and for solids and liquids, we can write

$$c \text{ grad } T = \text{grad } h , \qquad (2.6)$$

where c is the constant-pressure specific heat. With this substitution, the energy equation becomes

$$\text{div } (\rho \mathbf{u} h) = \text{div } \left(\frac{k}{c} \text{ grad } h \right) + S_h . \qquad (2.7)$$

If c is constant, the $h \sim T$ relation simplifies to

$$h = cT , \qquad (2.8)$$

which would lead to

$$\text{div } (\rho \mathbf{u} T) = \text{div } \left(\frac{k}{c} \text{ grad } T \right) + \frac{S_h}{c} . \qquad (2.9)$$

In this manner, either the enthalpy or the temperature can be chosen as the dependent variable.

The steady heat-conduction situation is obtained by setting the velocity $\mathbf{u}$ to zero; thus,

$$\text{div } (k \text{ grad } T) + S_h = 0 . \qquad (2.10)$$

2.1-4 A Momentum Equation

The differential equation governing the conservation of momentum in a given direction for a Newtonian fluid can be written along similar lines; however, the complication is greater because both shear and normal stresses must be considered and because the Stokes viscosity law is more complicated than Fick's law or Fourier's law. With u denoting the x-direction velocity, we write the corresponding momentum equation as

$$\frac{\partial}{\partial t} (\rho u) + \text{div } (\rho \mathbf{u} u) = \text{div } (\mu \text{ grad } u) - \frac{\partial p}{\partial x} + B_x + V_x , \qquad (2.11)$$

where μ is the viscosity, p is the pressure, B_x is the x-direction body force per unit volume, and V_x stands for the viscous terms that are in addition to those expressed by div $(\mu$ grad $u)$.

2.1-5 The Time-averaged Equations for Turbulent Flow

Turbulent flows are commonly encountered in practical applications. It is the time-mean behavior of these flows that is usually of practical interest.

Therefore, the equations for unsteady laminar flow are converted into the time-averaged equations for turbulent flow by an averaging operation in which it is assumed that there are rapid and random fluctuations about the mean value. The additional terms arising from this operation are the so-called Reynolds stresses, turbulent heat flux, turbulent diffusion flux, etc. To express these fluxes in terms of the mean properties of the flow is the task of a turbulence model.

Many turbulence models employ the concept of a turbulent viscosity or a turbulent diffusivity to express the turbulent stresses and fluxes. The result is that the time-averaged equations for turbulent flow have the same appearance as the equations for laminar flow, but the laminar exchange coefficients such as viscosity, diffusivity, and conductivity are replaced by *effective* (i.e., laminar plus turbulent) exchange coefficients. From a computational viewpoint, a turbulent flow within this framework is equivalent to a laminar flow with a rather complicated prescription of viscosity. (The same idea is applicable to non-Newtonian flows, which can be thought of as flows in which the viscosity depends on the velocity gradient.)

2.1-6 The Turbulence–Kinetic-Energy Equation

The currently popular "two-equation models" of turbulence (Launder and Spalding, 1972, 1974) employ, as one of the equations, the equation for the kinetic energy k of the fluctuating motion, which reads

$$\frac{\partial}{\partial t} (\rho k) + \text{div} (\rho \mathbf{u} k) = \text{div} (\Gamma_k \text{ grad } k) + G - \rho \epsilon , \qquad (2.12)$$

where Γ_k is the diffusion coefficient for k, G is the rate of generation of turbulence energy, and ϵ is the kinematic rate of dissipation. The quantity $G - \rho \epsilon$ is the *net* source term in the equation. A similar differential equation governs the variable ϵ.

2.1-7 The General Differential Equation

This brief journey through some of the relevant differential equations has indicated that all the dependent variables of interest here seem to obey a generalized conservation principle. If the dependent variable is denoted by ϕ, the general differential equation is

$$\frac{\partial}{\partial t} (\rho \phi) + \text{div} (\rho \mathbf{u} \phi) = \text{div} (\Gamma \text{ grad } \phi) + S , \qquad (2.13)$$

where Γ is the diffusion coefficient, and S is the source term. The quantities Γ and S are specific to a particular meaning of ϕ. (Indeed, we should have used the symbols Γ_ϕ and S_ϕ; this would, however, lead to too many subscripts in subsequent work.)

The four terms in the general differential equation are the unsteady term, the convection term, the diffusion term, and the source term. The dependent variable ϕ can stand for a variety of different quantities, such as the mass fraction of a chemical species, the enthalpy or the temperature, a velocity component, the turbulence kinetic energy, or a turbulence length scale. Accordingly, for each of these variables, an appropriate meaning will have to be given to the diffusion coefficient Γ and the source term S.

Not all diffusion fluxes are governed by the gradient of the relevant variable. The use of div (Γ grad ϕ) as the diffusion term does not, however, limit the general ϕ equation to gradient-driven diffusion processes. Whatever cannot be fitted into the *nominal* diffusion term can always be expressed as a part of the source term; in fact, the diffusion coefficient Γ can even be set equal to zero if desired. A gradient-diffusion term has been explicitly included in the general ϕ equation because most dependent variables do require a prominent diffusion term of this nature.

The density appearing in Eq. (2.13) may be related, via an equation of state, to variables such as mass fraction and temperature. These variables and the velocity components obey the general differential equation. Further, the flow field should satisfy an additional constraint, namely, the mass-conservation or the continuity equation, which is

$$\frac{\partial \rho}{\partial t} + \text{div} (\rho \mathbf{u}) = 0 \; . \tag{2.14}$$

We have written Eqs. (2.13) and (2.14) in vector form. Another useful representation is the Cartesian-tensor form of these equations:

$$\frac{\partial}{\partial t} (\rho \phi) + \frac{\partial}{\partial x_j} (\rho u_j \phi) = \frac{\partial}{\partial x_j} \left(\Gamma \frac{\partial \phi}{\partial x_j} \right) + S \tag{2.15}$$

$$\frac{\partial \rho}{\partial t} + \frac{\partial}{\partial x_j} (\rho u_j) = 0 \; , \tag{2.16}$$

where the subscript j can take the values 1, 2, 3, denoting the three space coordinates. When a subscript is repeated in a term, a summation of three terms is implied; for example,

$$\frac{\partial}{\partial x_j} (\rho u_j) = \frac{\partial}{\partial x_1} (\rho u_1) + \frac{\partial}{\partial x_2} (\rho u_2) + \frac{\partial}{\partial x_3} (\rho u_3) \tag{2.17}$$

$$\frac{\partial}{\partial x_j} \left(\Gamma \frac{\partial \phi}{\partial x_j} \right) = \frac{\partial}{\partial x_1} \left(\Gamma \frac{\partial \phi}{\partial x_1} \right) + \frac{\partial}{\partial x_2} \left(\Gamma \frac{\partial \phi}{\partial x_2} \right) + \frac{\partial}{\partial x_3} \left(\Gamma \frac{\partial \phi}{\partial x_3} \right) . \tag{2.18}$$

An immediate benefit of the Cartesian-tensor form is that the *one-dimensional* form of the equation is obtained by simply dropping the subscript j.

The procedure for casting any particular differential equation into the general form (2.13) is to manipulate it until, for the chosen dependent variable, the unsteady term and the convection and diffusion terms conform to the standard form. The coefficient of grad ϕ in the diffusion term is then taken as the expression for Γ, and the remaining terms on the right-hand side are collectively defined as the source term S.

Although we have so far considered all the variables as *dimensional* quantities, it is at times more convenient to work with *dimensionless* variables. Again, any particular differential equation written in terms of dimensionless variables can be regarded as possessing the general form (2.13), with ϕ standing for the dimensionless dependent variable, and with Γ and S being the dimensionless forms of the diffusion coefficient and the source term. In many cases, the dimensionless value of Γ may simply be unity, while S may take the value of 0 or 1.

The recognition that all the relevant differential equations for heat and mass transfer, fluid flow, turbulence, and related phenomena can be thought of as particular cases of the general ϕ equation is an important time-saving step. As a consequence, we need to concern ourselves with the numerical solution of only Eq. (2.13). Even in the construction of a computer program, it is sufficient to write a general sequence of instructions for solving Eq. (2.13), which can be repeatedly used for different meanings of ϕ along with appropriate expressions for Γ and S, and, of course, with appropriate initial and boundary conditions. Thus, the concept of the general ϕ equation enables us to formulate a general numerical method and to prepare general-purpose computer programs.

2.2 NATURE OF COORDINATES

So far we have given attention to the dependent variables. Now we shall turn to the independent variables and discuss their properties from the computational point of view.

2.2-1 Independent Variables

The dependent variable ϕ would, in general, be a function of three space coordinates and time. Thus,

$$\phi = \phi(x, y, z, t) \, , \tag{2.19}$$

where x, y, z, and t are the *independent* variables. In a numerical solution, we shall *choose* the values of the independent variables at which the values of ϕ are to be calculated.

Fortunately, not all problems require consideration of all four independent variables. The smaller the number of participating independent variables, the fewer will be the locations (or grid points) at which the ϕ values must be calculated (provided that otherwise the problems are of comparable complexity).

When the relevant physical quantities depend on only one space coordinate, the situation is called *one-dimensional*. Dependence on two space coordinates leads to a *two-dimensional* situation, and on three space coordinates to a *three-dimensional* situation. When the problem contains no dependence on time, it is called *steady*. Otherwise, it is called *unsteady* or time-dependent. Considering the dependence on space and time together, we shall describe a situation as an unsteady one-dimensional problem, a steady three-dimensional flow, etc.

The choice of independent coordinates as expressed by Eq. (2.19) is not the only possibility. Instead of describing a steady temperature distribution as $T(x, y, z)$, we may write

$$z = z(T, x, y) , \qquad (2.20)$$

where z becomes the dependent variable that stands for the height of an isothermal surface corresponding to T at the location (x, y). A method based on such a representation has been developed by Dix and Cizek (1970) and by Crank and co-workers (Crank and Phahle, 1973; Crank and Gupta, 1975; Crank and Crowley, 1978) and is known as the *isotherm migration* method. The method is, however, limited to temperature fields that are monotonic functions of the coordinates; for more general fields, the height z could have several values for given values of T, x, and y; this makes z, for computational purposes, unsuitable as a dependent variable.

2.2-2 Proper Choice of Coordinates

Since the number of grid points would, in general, be related to the number of independent variables, there is a significant computational saving to be achieved by working with fewer independent variables. A judicious choice of the coordinate system can sometimes reduce the number of independent variables required.

Although we have used x, y, and z as the space coordinates, it is not implied that we must use the Cartesian coordinate system; any description of the spatial location will do. We shall now illustrate, by a few specific examples, how the choice of coordinates influences the number of independent variables.

1. The flow around an airplane that is moving with constant velocity is *unsteady* when viewed from a stationary coordinate system, but *steady* with respect to a moving coordinate system attached to the airplane.

2. The axisymmetric flow in a circular pipe appears to be three-dimensional in a Cartesian coordinate system but is two-dimensional in cylindrical polar coordinates r, θ, z, since

$$\phi = \phi(r, z) \tag{2.21}$$

with no dependence on θ.

3. Transformed coordinates offer further possibilities of fewer independent variables. For example:

   a. A two-dimensional laminar boundary layer on a flat plate gives a similarity behavior such that the velocity u depends on η alone, where

$$\eta = \frac{cy}{\sqrt{x}} \tag{2.22}$$

   and where c is a dimensional constant. Thus, a two-dimensional problem is reduced to a one-dimensional problem.

   b. Unsteady heat conduction in a semi-infinite solid has x and t as the independent variables. However, for some simple boundary conditions, the temperature can be shown to depend on ξ alone, where

$$\xi = \frac{Cx}{\sqrt{t}} \ , \tag{2.23}$$

   with C representing an appropriate dimensional constant.

4. A change of the dependent variable can lead to a reduction in the number of independent variables. For example:

   a. In a fully developed duct flow, the temperature T depends on the streamwise coordinate x and the cross-stream coordinate y. However, in the thermally developed regime with uniform wall temperature T_w, we have

$$\theta = \theta(y), \tag{2.24}$$

   where

$$\theta \equiv \frac{T - T_w}{T_b - T_w}$$

   and T_b is the bulk temperature, which varies with x.

   b. A plane free jet is a two-dimensional flow. However, we can write

$$\tilde{u} = \tilde{u}(\eta), \tag{2.25}$$

where

$$\tilde{u} \equiv \frac{u}{u_c}, \qquad \eta \equiv \frac{y}{\delta}. \tag{2.26}$$

Here u_c represents the center-line velocity, y is the cross-stream coordinate, and δ is a characteristic jet width. Both u_c and δ vary with the streamwise coordinate x.

Although most of the discussion in this book will be conducted in terms of x, y, z, and t as the independent variables, it should be remembered that all the ideas and practices are equally applicable to the transformed or dimensionless variables illustrated here. Indeed, for computational efficiency, numerical methods should always be used with the appropriate choice of coordinates.

2.2-3 One-Way and Two-Way Coordinates

We shall now consider new concepts about the properties of coordinates and then establish a connection between these and the standard mathematical terminology.

Definitions. A *two-way* coordinate is such that the conditions at a given location in that coordinate are influenced by changes in conditions on *either side* of that location. A *one-way* coordinate is such that the conditions at a given location in that coordinate are influenced by changes in conditions on *only one side* of that location.

Examples. One-dimensional steady heat conduction in a rod provides an example of a two-way coordinate. The temperature of any given point in the rod can be influenced by changing the temperature of either end. Normally, space coordinates are two-way coordinates. Time, on the other hand, is *always* a one-way coordinate. During the unsteady cooling of a solid, the temperature at a given instant can be influenced by changing only those conditions that prevailed *before* that instant. It is a matter of common experience that yesterday's events affect today's happenings, but tomorrow's conditions have no influence on what happens today.

Space as a one-way coordinate. What is more interesting is that even a space coordinate can very nearly become one-way under the action of fluid flow. If there is a strong unidirectional flow in the coordinate direction, then significant influences travel only from upstream to downstream. The conditions at a point are then affected largely by the upstream conditions, and very little by the downstream ones. The one-way nature of a space coordinate is an approximation. It is true that convection is a one-way process, but diffusion (which is always present) has two-way influences. However, when the flow rate is large, convection overpowers diffusion and thus makes the space coordinate nearly one-way.

Parabolic, elliptic, hyperbolic. It appears that the mathematical terms *parabolic* and *elliptic*, which are used for the classification of differential equations, correspond to our computational concepts of one-way and two-way coordinates. The term parabolic indicates a one-way behavior, while elliptic signifies the two-way concept.

It would be more meaningful if situations were described as being parabolic or elliptic in a given coordinate. Thus, the unsteady heat conduction problem, which is normally called parabolic, is actually parabolic in time and elliptic in the space coordinates. The steady heat conduction problem is elliptic in all coordinates. A two-dimensional boundary layer is parabolic in the streamwise coordinate and elliptic in the cross-stream coordinate.

Since such descriptions are unconventional, a connection with established practice can perhaps be achieved by the following rule:

A situation is parabolic if there exists at least one one-way coordinate; otherwise, it is elliptic.

A flow with one one-way space coordinate is sometimes called a boundary-layer-type flow, while a flow with all two-way coordinates is referred to as a recirculating flow [see the titles of the books by Patankar and Spalding (1970) and Gosman, Pun, Runchal, Spalding, and Wolfshtein (1969)].

What about the third category, namely, hyperbolic? It so happens that a hyperbolic situation does not neatly fit into the computational classification. A hyperbolic problem has a kind of one-way behavior, which is, however, not along coordinate directions but along special lines called *characteristics.* There are numerical methods that make use of the characteristic lines, but they are restricted to hyperbolic problems. On the other hand, the numerical method to be developed in this book does not take advantage of the special nature of a hyperbolic problem. We shall treat hyperbolic problems as members of the general class of elliptic problems (i.e., all two-way coordinates).

Computational implications. The motivation for the foregoing discussion about one-way and two-way coordinates is that, if a one-way coordinate can be identified in a given situation, substantial economy of computer storage and computer time is possible. Let us consider an unsteady two-dimensional heat conduction problem. We shall construct a two-dimensional array of grid points in the calculation domain. At any instant of time, there will be a corresponding two-dimensional temperature field. Such a field will have to be handled in the computer for each of the successive instants of time. However, since time is a one-way coordinate, the temperature field at a given time is not affected by the *future* temperature fields. Indeed, the entire unsteady problem can be reduced to the required repetitions of one basic step, namely this: Given the temperature field at time t, find the temperature field at time $t + \Delta t$. Thus, computer storage will be needed only for these two temperature fields; the same storage space can be used, over and over again, for all the time steps.

In this manner, starting with a given *initial* temperature field, we are able to "march" forward to successive instants of time. During any time step, only one two-dimensional array of temperatures forms the unknowns to be treated simultaneously.[*] They are decoupled from all future values of temperature, and the previous values that influence them are known. Thus, we need to solve a much simpler set of equations, with a consequent saving of computer time.

In a similar manner, a two-dimensional boundary layer is computed by marching in the streamwise coordinate. With values of the dependent variables given along one cross-stream line at an upstream station, the values along successive cross-stream lines are obtained. Only one-dimensional computer storage is needed for handling the two-dimensional flow. Similarly, a three-dimensional duct flow that is parabolic in the streamwise direction can be treated as a series of two-dimensional problems for successive cross-stream planes.

In this book, we shall give only occasional attention to the one-way *space* coordinate. However, its great potential for saving computer storage and computer time should always be kept in mind.

PROBLEMS

2.1 Write the unsteady heat conduction equation for the case of constant specific heat c. Show that, with reference to the general equation (2.13), this implies $\phi = T$, $\mathbf{u} = 0$, $\Gamma = k/c$, and $S = S_h/c$.

2.2 Derive the expressions for ϕ, Γ, and S if in Problem 2.1 the specific heat c cannot be taken as constant. (*Hint*: Use the internal energy i as the dependent variable; note that $di = c\,dT$.)

2.3 If Eq. (2.7) were to be written for an *unsteady* situation, show that the resulting form can be expressed as $\phi = h$, $\Gamma = k/c$, and $S = S_h + \partial p/\partial t$.

2.4 Derive an expression for V_x in Eq. (2.11). Hence show that V_x becomes zero when the density and viscosity are constant. (Use the continuity equation.)

2.5 Define an effective pressure by

$$P = p - \tfrac{1}{3}\mu \text{ div } \mathbf{u} \, ,$$

where p is the thermodynamic pressure. If the viscosity is constant but the density ρ is *not* constant (and hence div $\mathbf{u} \neq 0$), show that the term V_x in Eq. (2.11) can be combined with the pressure gradient such that

$$-\frac{\partial p}{\partial x} + V_x = -\frac{\partial P}{\partial x} \, .$$

---

[*]It is assumed here that an *implicit* method is to be employed. This matter is discussed in detail in Chapter 4.

2.6 If the continuity equation (2.14) were to be regarded as a special case of the general equation (2.13), what would be the expressions for ϕ, Γ, and S?

2.7 Consider a mixture of various chemical species. Define the mixture enthalpy by $h = \Sigma\, m_l h_l$, where m_l is the mass fraction of a typical species, and h_l is its specific enthalpy, which is given by

$$h_l = h_l^0 + \int_0^T c_l\, dT \ .$$

Here h_l^0 is a constant, and c_l is the constant-pressure specific heat of species l. Write the steady-state enthalpy-conservation equation and hence show that $\phi = h$, $\Gamma = k/c$, and $S = S_h + \mathrm{div}\ \Sigma\ [(\Gamma_l - k/c)h_l\ \mathrm{grad}\ m_l]$, where c is the mixture specific heat, given by $\Sigma\, m_l c_l$.

THREE

DISCRETIZATION METHODS

So far we have seen that there are significant benefits in obtaining a theoretical prediction of physical phenomena. The phenomena of interest here are governed by differential equations, which we have represented by a general equation for the variable ϕ. Now our main task is to develop the means of solving this equation.

For ease of understanding, we shall assume in this chapter that the variable ϕ is a function of only one independent variable x. However, the ideas developed here continue to be applicable when more than one independent variable is active.

3.1 THE NATURE OF NUMERICAL METHODS

3.1-1 The Task

A numerical solution of a differential equation consists of a set of numbers from which the distribution of the dependent variable ϕ can be constructed. In this sense, a numerical method is akin to a laboratory experiment, in which a set of instrument readings enables us to establish the distribution of the measured quantity in the domain under investigation. The numerical analyst and the laboratory experimenter both must remain content with only a *finite* number of numerical values as the outcome, although this number can, at least in principle, be made large enough for practical purposes.

Let us suppose that we decide to represent the variation of ϕ by a polynomial in x,

$$\phi = a_0 + a_1 x + a_2 x^2 + \cdots + a_m x^m , \tag{3.1}$$

and employ a numerical method to find the finite number of coefficients a_0, a_1, a_2, $\ldots$, a_m. This will enable us to evaluate ϕ at any location x by substituting the value of x and the values of the a's into Eq. (3.1). This procedure is, however, somewhat inconvenient if our ultimate interest is to obtain the *values* of ϕ at various locations. The values of the a's are, by themselves, not particularly meaningful, and the substitution operation must be carried out to arrive at the required values of ϕ. This leads us to the following thought: Why not construct a method that employs the *values* of ϕ at a number of given points as the primary unknowns? Indeed, most numerical methods for solving differential equations do belong in this category, and therefore we shall limit our attention to such methods.

Thus, a numerical method treats as its basic unknowns the values of the dependent variable at a finite number of locations (called the *grid points*) in the calculation domain. The method includes the tasks of providing a set of algebraic equations for these unknowns and of prescribing an algorithm for solving the equations.

3.1-2 The Discretization Concept

In focusing attention on the values at the grid points, we have replaced the continuous information contained in the exact solution of the differential equation with discrete values. We have thus discretized the distribution of ϕ, and it is appropriate to refer to this class of numerical methods as *discretization methods*.

The algebraic equations involving the unknown values of ϕ at chosen grid points, which we shall now name the *discretization equations*, are derived from the differential equation governing ϕ. In this derivation, we must employ some assumption about how ϕ varies *between* the grid points. Although this "profile" of ϕ could be chosen such that a single algebraic expression suffices for the whole calculation domain, it is often more practical to use *piecewise* profiles such that a given segment describes the variation of ϕ over only a small region in terms of the ϕ values at the grid points within and around that region. Thus, it is common to subdivide the calculation domain into a number of subdomains or elements such that a separate profile assumption can be associated with each subdomain.

In this manner, we encounter the discretization concept in another context. The continuum calculation domain has been discretized. It is this systematic discretization of space and of the dependent variables that makes it possible to replace the governing differential equations with simple algebraic equations, which can be solved with relative ease.

3.1-3 The Structure of the Discretization Equation

A discretization equation is an algebraic relation connecting the values of ϕ for a group of grid points. Such an equation is derived from the differential equation governing ϕ and thus expresses the same physical information as the differential equation. That only a few grid points participate in a given discretization equation is a consequence of the piecewise nature of the profiles chosen. The value of ϕ at a grid point thereby influences the distribution of ϕ only in its immediate neighborhood. As the number of grid points becomes very large, the solution of the discretization equations is expected to approach the exact solution of the corresponding differential equation. This follows from the consideration that, as the grid points get closer together, the change in ϕ between neighboring grid points becomes small, and then the actual details of the profile assumption become unimportant.

For a given differential equation, the possible discretization equations are by no means unique, although all types of discretization equations are, in the limit of a very large number of grid points, expected to give the same solution. The different types arise from the differences in the profile assumptions and in the methods of derivation.

Until now we have deliberately refrained from making reference to finite-difference and finite-element methods. Now it may be stated that these can be thought of as two alternative versions of the discretization method, which we have described in general terms. The distinction between the finite-difference method and the finite-element method results from the ways of choosing the profiles and deriving the discretization equations. The method that is to be the main focus of attention in this book has the *appearance* of a finite-difference method, but it employs many ideas that are typical of the finite-element methodology. To call the present method a finite-difference method might convey an adherence to the conventional finite-difference practice. For this reason, we shall refer to it simply as a discretization method. Also, we shall note in Chapter 8 how a method that has the appearance of a finite-element method can be constructed from the general principles presented in this book.

3.2 METHODS OF DERIVING THE DISCRETIZATION EQUATIONS

For a given differential equation, the required discretization equations can be derived in many ways. Here, we shall outline a few common methods and then indicate a preference.

3.2-1 Taylor-Series Formulation

The usual procedure for deriving finite-difference equations consists of approximating the derivatives in the differential equation via a truncated Taylor series. Let us consider the grid points shown in Fig. 3.1. For grid point 2, located midway between grid points 1 and 3 such that $\Delta x = x_2 - x_1 = x_3 - x_2$, the Taylor-series expansion around 2 gives

$$\phi_1 = \phi_2 - \Delta x \left(\frac{d\phi}{dx}\right)_2 + \frac{1}{2} (\Delta x)^2 \left(\frac{d^2\phi}{dx^2}\right)_2 - \cdots \qquad (3.2)$$

and

$$\phi_3 = \phi_2 + \Delta x \left(\frac{d\phi}{dx}\right)_2 + \frac{1}{2} (\Delta x)^2 \left(\frac{d^2\phi}{dx^2}\right)_2 + \cdots . \qquad (3.3)$$

Truncating the series just after the third term, and adding and subtracting the two equations, we obtain

$$\left(\frac{d\phi}{dx}\right)_2 = \frac{\phi_3 - \phi_1}{2 \Delta x} \qquad (3.4)$$

and

$$\left(\frac{d^2\phi}{dx^2}\right)_2 = \frac{\phi_1 + \phi_3 - 2\phi_2}{(\Delta x)^2} . \qquad (3.5)$$

The substitution of such expressions into the differential equation leads to the finite-difference equation.

The method includes the assumption that the variation of ϕ is somewhat like a polynomial in x, so that the higher derivatives are unimportant. This assumption, however, leads to an undesirable formulation when, for example, exponential variations are encountered. (We shall refer to this matter again in Chapter 5.) The Taylor-series formulation is relatively straightforward but allows less flexibility and provides little insight into the physical meanings of the terms.[*]

---

[*]This is admittedly an entirely subjective view. Someone with proper mathematical training may find the Taylor-series method highly illuminating and meaningful.

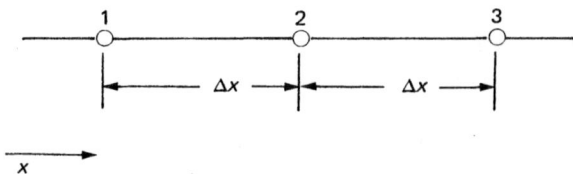

Figure 3.1 Three successive grid points used for the Taylor-series expansion.

3.2-2 Variational Formulation

Another method of obtaining the discretization equations is based on the calculus of variations. To understand the method fully, the reader should have sufficient knowledge of this branch of calculus. However, a general appreciation of the main ingredients of the formulation is all that is needed for the present purposes.

The calculus of variations shows that solving certain differential equations is equivalent to minimizing a related quantity called the *functional*. This equivalence is known as a variational principle. If the functional is minimized with respect to the grid-point values of the dependent variable, the resulting conditions give the required discretization equations. The variational formulation is very commonly employed in finite-element methods for stress analysis, where it can be linked to the virtual-work principle. In addition to its algebraic and conceptual complexity, the main drawback of this formulation is its limited applicability, since a variational principle does not exist for all differential equations of interest.

3.2-3 Method of Weighted Residuals

A powerful method for solving differential equations is the method of weighted residuals, which is described in detail by Finlayson (1972). The basic concept is simple and interesting. Let the differential equation be represented by

$$L(\phi) = 0 . \tag{3.6}$$

Further, let us assume an approximate solution $\bar{\phi}$ that contains a number of undetermined parameters, for example,

$$\bar{\phi} = a_0 + a_1 x + a_2 x^2 + \cdots + a_m x^m , \tag{3.7}$$

the a's being the parameters. The substitution of $\bar{\phi}$ into the differential equation leaves a residual R, defined as

$$R = L(\bar{\phi}) . \tag{3.8}$$

We wish to make this residual small in some sense. Let us propose that

$$\int WR \, dx = 0 , \tag{3.9}$$

where W is a weighting function and the integration is performed over the domain of interest. By choosing a succession of weighting functions, we can generate as many equations as are required for evaluating the parameters. These algebraic equations containing the parameters as the unknowns are solved to obtain the approximate solution to the differential equation. Different versions of the method (known by specific names) result from the choice of different classes of weighting functions.

The method was very popular in boundary-layer analysis before the finite-difference method nearly replaced it. However, a connection with the finite-difference method, or rather with the discretization method, can be established if the approximate solution $\bar{\phi}$, instead of being a single algebraic expression over the whole domain, is constructed via piecewise profiles with the grid-point values of ϕ as the unknown parameters. Indeed, much of the recent development of the finite-element technique is also based on piecewise profiles used in conjunction with a particular weighted-residual practice known as the Galerkin method.

The simplest weighting function is $W = 1$. From this, a number of weighted-residual equations can be generated by dividing the calculation domain into subdomains or control volumes, and setting the weighting function to be unity over one subdomain at a time and zero everywhere else. This variant of the method of weighted residuals is called the *subdomain* method or the *control-volume* formulation. It implies that the integral of the residual over each control volume must become zero. Since we shall adopt the control-volume approach in this book, a more detailed discussion is desirable, which now follows.

3.2-4 Control-Volume Formulation

Often elementary textbooks on heat transfer derive the finite-difference equation via the Taylor-series method and then demonstrate that the resulting equation is consistent with a heat balance over a small region surrounding a grid point. We have also seen that the control-volume formulation can be regarded as a special version of the method of weighted residuals. The basic idea of the control-volume formulation is easy to understand and lends itself to direct physical interpretation. The calculation domain is divided into a number of nonoverlapping control volumes such that there is one control volume surrounding each grid point. The differential equation is integrated over each control volume. Piecewise profiles expressing the variation of ϕ between the grid points are used to evaluate the required integrals. The result is the discretization equation containing the values of ϕ for a group of grid points.

The discretization equation obtained in this manner expresses the conservation principle for ϕ for the finite control volume, just as the differential equation expresses it for an infinitesimal control volume.[*]

---

[*]Indeed, deriving the control-volume discretization equation by integrating the differential equation over a finite control volume is a rather roundabout process, much

The most attractive feature of the control-volume formulation is that the resulting solution would imply that the *integral* conservation of quantities such as mass, momentum, and energy is exactly satisfied over any group of control volumes and, of course, over the whole calculation domain. This characteristic exists for *any* number of grid points—not just in a limiting sense when the number of grid points becomes large. Thus, even the coarse-grid solution exhibits *exact* integral balances.

When the discretization equations are solved to obtain the grid-point values of the dependent variable, the result can be viewed in two different ways. In the finite-element method and in most weighted-residual methods, the assumed variation of ϕ consisting of the grid-point values and the interpolation functions (or profiles) between the grid points is taken as the approximate solution. In the finite-difference method, however, only the grid-point values of ϕ are considered to constitute the solution, without any explicit reference as to how ϕ varies between the grid points. This is akin to a laboratory experiment where the distribution of a quantity is obtained in terms of the measured values at some discrete locations without any statement about the variation *between* these locations. In our control-volume approach, we shall also adopt this view. We shall seek the solution in the form of the grid-point values only. The interpolation formulas or the profiles will be regarded as auxiliary relations needed to evaluate the required integrals in the formulation. Once the discretization equations are derived, the profile assumptions can be forgotten. This viewpoint permits complete freedom of choice in employing, if we wish, different profile assumptions for integrating different terms in the differential equation.

To make the foregoing discussion more concrete, we shall now derive the control-volume discretization equation for a simple situation.

3.3 AN ILLUSTRATIVE EXAMPLE

Let us consider steady one-dimensional heat conduction governed by

$$\frac{d}{dx}\left(k\,\frac{dT}{dx}\right) + S = 0\ ,\qquad (3.10)$$

where k is the thermal conductivity, T is the temperature, and S is the rate of heat generation per unit volume.

---

like preparing mashed potatoes from dehydrated potato powder. After all, textbook derivations of differential equations always start from the conservation principle applied to a small control volume. It is useful to imagine ourselves to be in the pre-calculus days; then the control-volume equation would have been our only way of stating the conservation principle.

Preparation. To derive the discretization equation, we shall employ the grid-point cluster shown in Fig. 3.2. We focus attention on the grid point P, which has the grid points E and W as its neighbors. (E denotes the east side, i.e., the positive x direction, while W stands for west or the negative x direction.) The dashed lines show the faces of the control volume; their exact locations are unimportant for the time being. The letters e and w denote these faces. For the one-dimensional problem under consideration, we shall assume a unit thickness in the y and z directions. Thus, the volume of the control volume shown is $\Delta x \times 1 \times 1$. If we integrate Eq. (3.10) over the control volume, we get

$$\left(k\frac{dT}{dx}\right)_e - \left(k\frac{dT}{dx}\right)_w + \int_w^e S\,dx = 0 . \tag{3.11}$$

Profile assumption. To make further progress, we need a profile assumption or an interpolation formula. Two simple profile assumptions are shown in Fig. 3.3. The simplest possibility is to assume that the value of T at a grid point prevails over the control volume surrounding it. This gives the stepwise profile sketched in Fig. 3.3a. For this profile, the slope dT/dx is not defined at the control-volume faces (i.e., at w or e). A profile that does not suffer from this difficulty is the piecewise-linear profile (Fig. 3.3b). Here, linear interpolation functions are used between the grid points.

The discretization equation. If we evaluate the derivatives dT/dx in Eq. (3.11) from the piecewise-linear profile, the resulting equation will be

$$\frac{k_e(T_E - T_P)}{(\delta x)_e} - \frac{k_w(T_P - T_W)}{(\delta x)_w} + \bar{S}\,\Delta x = 0 , \tag{3.12}$$

where $\bar{S}$ is the average value of S over the control volume. It is useful to cast the discretization equation (3.12) into the following form:

$$a_P T_P = a_E T_E + a_W T_W + b , \tag{3.13}$$

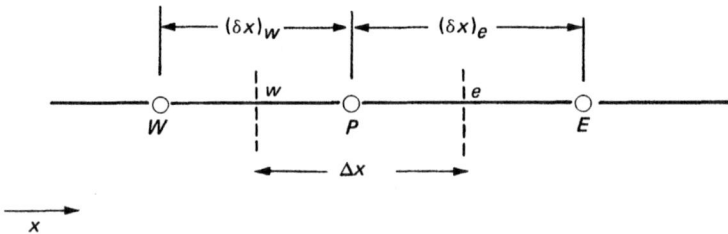

Figure 3.2 Grid-point cluster for the one-dimensional problem.

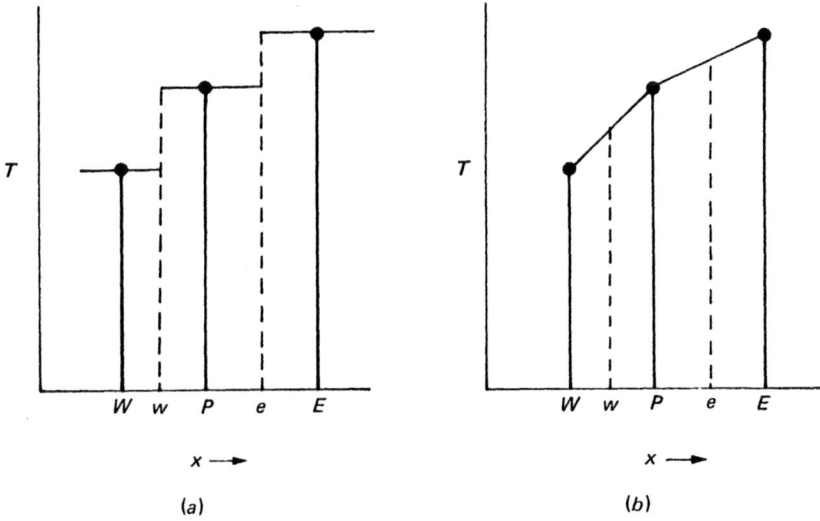

Figure 3.3 Two simple profile assumptions. (*a*) Stepwise profile; (*b*) piecewise-linear profile.

where

$$a_E = \frac{k_e}{(\delta x)_e} , \qquad (3.14a)$$

$$a_W = \frac{k_w}{(\delta x)_w} , \qquad (3.14b)$$

$$a_P = a_E + a_W , \qquad (3.14c)$$

and
$$b = \bar{S} \, \Delta x . \qquad (3.14d)$$

Comments.

1. Equation (3.13) represents the standard form in which we shall write our discretization equations. The temperature T_P at the central grid point appears on the left side of the equation, while the neighbor-point temperatures and the constant b form the terms on the right side. As we shall see later, the number of neighbors increases for two- and three-dimensional situations. In general, it is convenient to think of Eq. (3.13) as having the form

$$a_P T_P = \Sigma \, a_{nb} T_{nb} + b , \qquad (3.15)$$

where the subscript nb denotes a neighbor, and the summation is to be taken over all the neighbors.

2. In deriving Eq. (3.13), we have used the simplest profile assumption that enabled us to evaluate dT/dx. Of course, many other interpolation functions would have been possible.

3. Further, it is important to understand that we need not use the same profile for all quantities. For example, $\bar{S}$ need not be calculated from a linear variation of S between the grid points, nor k_e from a linear variation of k between k_P and k_E.

4. Even for a given variable, the same profile assumption need not be used for all terms in the equation. For example, if Eq. (3.10) had an additional term involving T alone, it would have been permissible to use a stepwise profile for that term, instead of adhering to the piecewise-linear profile used for evaluating dT/dx.

Guiding principles. The freedom of choice indicated so far gives rise to a variety of discretization formulations. It is true that, as the number of grid points is increased, all the formulations are expected to give the same solution. We shall, however, impose an additional requirement that will enable us to narrow down the number of acceptable formulations. We shall require that even the coarse-grid solution should always have (1) physically realistic behavior and (2) overall balance.

Physical realism is easy to understand, at least in simple cases. The variations shown in Fig. 3.4 illustrate this concept. A realistic variation should have the same qualitative trend as the exact variation. In heat conduction without sources, no temperature can lie outside the range of temperature established by the boundary temperatures. When a hot solid is being cooled

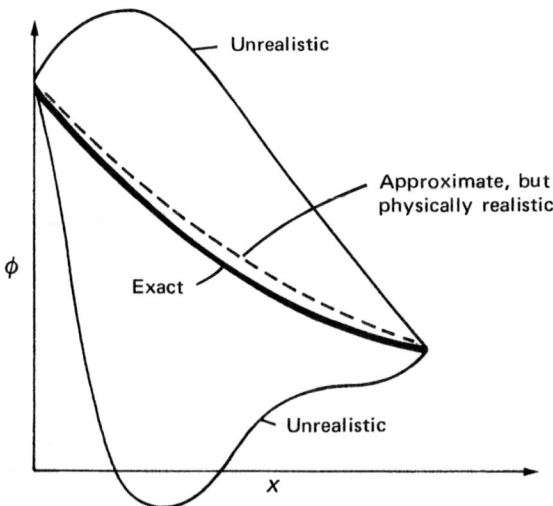

Figure 3.4 Physically realistic and unrealistic behavior.

by an ambient fluid, the solid cannot acquire a temperature lower than that of the fluid. We shall always apply such tests to our discretization equations.

The requirement of overall balance implies integral conservation over the whole calculation domain. We shall insist that the heat fluxes, mass flow rates, and momentum fluxes must correctly give an overall balance with appropriate sources and sinks—not just in the limit as the number of grid points becomes very large, but for any number of grid points. Our control-volume formulation makes this overall balance possible, but care is needed, as we shall shortly see, in calculating fluxes at the control-volume interfaces.

The constraints of physical realism and overall balance will be used to guide our choices of profile assumptions and related practices. On the basis of these constraints, we shall develop some basic rules that will enable us to discriminate between available formulations and to invent new ones. The decisions that are normally governed by mathematical considerations can now be directed by physical reasoning.

Treatment of the source term. Before we proceed to develop the basic rules, we shall give some attention to the source term S in Eq. (3.10). Often, the source term is a function of the dependent variable T itself, and it is then desirable to acknowledge this dependence in constructing the discretization equation. We can, however, formally account for only a linear dependence because, as we shall see later, the discretization equations will be solved by the techniques for linear algebraic equations. The procedure for "linearizing" a given $S \sim T$ relationship will be discussed in the next chapter. Here, it is sufficient to express the average value $\bar{S}$ as

$$\bar{S} = S_C + S_P T_P , \qquad (3.16)$$

where S_C stands for the constant part of $\bar{S}$, while S_P is the coefficient of T_P. (Obviously, S_P does *not* stand for S evaluated at point P.)

The appearance of T_P in Eq. (3.16) reveals that, in expressing the average value $\bar{S}$, we have presumed that the value T_P prevails over the control volume; in other words, the stepwise profile shown in Fig. 3.3a has been used. (It should be noted that we are free to use the stepwise profile for the source term while using the piecewise-linear profile for the dT/dx term.)

With the linearized source expression, the discretization equation would still look like Eq. (3.13), but the coefficient definitions [Eqs. (3.14)] would change. The new set is

$$a_P T_P = a_E T_E + a_W T_W + b , \qquad (3.17)$$

where

$$a_E = \frac{k_e}{(\delta x)_e} , \qquad (3.18a)$$

$$a_W = \frac{k_w}{(\delta x)_w} , \tag{3.18b}$$

$$a_P = a_E + a_W - S_P \, \Delta x , \tag{3.18c}$$

and $\qquad\qquad b = S_C \, \Delta x . \tag{3.18d}$

The foregoing introductory discussion provides sufficient background to allow the formulation of the basic rules that our discretization equations should obey, to ensure physical realism and overall balance. These seemingly simple rules have far-reaching implications, and they will guide the development of methods throughout this book.

3.4 THE FOUR BASIC RULES

Rule 1: Consistency at control-volume faces When a face is common to two adjacent control volumes, the flux across it must be represented by the *same* expression in the discretization equations for the two control volumes.

Discussion. Obviously, the heat flux that leaves one control volume through a particular face must be identical to the flux that enters the next control volume through the same face. Otherwise, the overall balance would not be satisfied. Although this requirement is easy to understand, subtle violations must be watched for. For the control volume shown in Fig. 3.2, we could have evaluated the interface heat fluxes $k \, dT/dx$ from a quadratic profile passing through T_W, T_P, and T_E. The use of the same kind of formulation for the next control volume implies that the gradient dT/dx at the common interface is calculated from different profiles, depending on which control volume is being considered. The resulting inconsistency[*] in dT/dx (and hence in the heat flux) is sketched in Fig. 3.5.

Another practice that could lead to flux inconsistency is to assume that the fluxes at the faces of a given control volume are all governed by the center-point conductivity k_P. Then the heat flux at the interface e (shown in Fig. 3.2) will be expressed as $k_P \, (T_P - T_E)/(\delta x)_e$ when the control volume surrounding the point P is considered, and as $k_E \, (T_P - T_E)/(\delta x)_e$ when the equation with E as the center point is constructed. To avoid such incon-

---

[*]It so happens that, if the interfaces are located *midway* between the grid points, the type of quadratic profile shown in Fig. 3.5 does not give any inconsistency. This is because the slope of a parabola at a location midway between two points is exactly equal to the slope of the straight line joining the two points. But this property of the parabola must be regarded as fortuitous, and one must, in general, refrain from changing the interface flux expression while going from one control volume to the next.

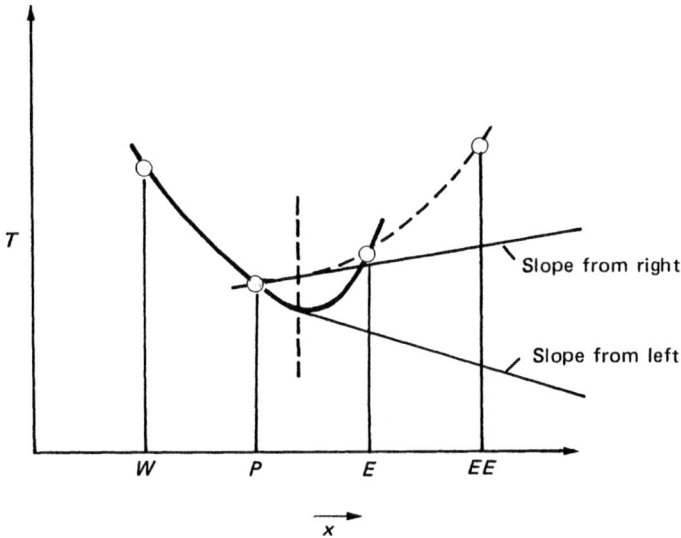

Figure 3.5 Flux inconsistency resulting from quadratic profile.

sistencies, it is useful to remember that an interface flux must be considered in its own right, and not as belonging to a certain control volume.

> **Rule 2: Positive coefficients** Most situations of interest here will be such that the value of a dependent variable at a grid point is influenced by the values at neighboring grid points only through the processes of convection and diffusion. Then it follows that an *increase* in the value at one grid point should, with other conditions remaining unchanged, lead to an *increase* (and *not* a decrease) in the value at the neighboring grid point. In Eq. (3.13), if an increase in T_E must lead to an increase in T_P, it follows that the coefficients a_E and a_P must have the same sign. In other words, for the general equation (3.15), the neighbor coefficients a_{nb} and the center-point coefficient a_P all must be of the same sign. We can, of course, choose to make them all positive or all negative. Let us decide to write our discretization equations such that the coefficients are positive; then Rule 2 can be stated as follows:
>
> > All coefficients (a_P and neighbor coefficients a_{nb}) must always be positive.

Comments. The coefficient definitions given in Eqs. (3.14) show that our illustrative discretization equation [Eq. (3.13)] does obey the positive-coefficient rule. However, as we shall see later, there are numerous formulations that frequently violate this rule. Usually, the consequence is a physically unrealistic solution. The presence of a negative neighbor coefficient can lead to the situation in which an increase in a boundary temperature causes the temperature at the adjacent grid point to decrease. We shall accept only those formulations that guarantee positive coefficients under all circumstances.

Rule 3: Negative-slope linearization of the source term If we consider the coefficient definitions in Eqs. (3.18), it appears that, even if the neighbor coefficients are positive, the center-point coefficient a_P can become negative via the S_P term. Of course, the danger can be completely avoided by requiring that S_P will not be positive. Thus, we formulate Rule 3 as follows:

When the source term is linearized as $\bar{S} = S_C + S_P T_P$, the coefficient S_P must always be less than or equal to zero.

Remarks. This rule is not as arbitrary as it sounds. Most physical processes do have a negative-slope relationship between the source term and the dependent variable. Indeed, if S_P were positive, the physical situation could become unstable. A positive S_P implies that, as T_P increases, the source term increases; if an effective heat-removal mechanism is not available, this may, in turn, lead to an increase in T_P, and so on. Computationally, it is vital to keep S_P negative so that instabilities and physically unrealistic solutions do not arise. The source-term linearization is further discussed in the next chapter. It is sufficient to note here that, for computational success, the principle of negative S_P is essential.

Rule 4: Sum of the neighbor coefficients Often the governing differential equations contain only the derivatives of the dependent variable. Then, if T represents the dependent variable, the functions T and $T + c$ (where c is an arbitrary constant) both satisfy the differential equation. This property of the differential equation must also be reflected by the discretization equation. Thus, Eq. (3.15) should remain valid even when T_P and all T_{nb}'s are increased by a constant. From this requirement, it follows that a_P must equal the sum of the neighbor coefficients. Hence, the statement of Rule 4 is:

We require

$$a_P = \Sigma \, a_{nb} \qquad (3.19)$$

for situations where the differential equation continues to remain satisfied after a constant is added to the dependent variable.

Discussion. It is easy to see that Eq. (3.13) does satisfy this rule. The rule implies that the center-point value T_P is a weighted average of the neighbor values T_{nb}. Unlike Eq. (3.13), the coefficients in Eq. (3.17) do not obey the rule. This is, however, not a violation, but a case of inapplicability of the rule. When the source term depends on T, both T and $T + c$ do not satisfy the differential equation. Even in such cases, the rule should not be forgotten, but should be applied by envisaging a special case of the equation. If, for example, S_P is set equal to zero in Eq. (3.17), the rule becomes applicable and is indeed obeyed.

When the differential equation is satisfied by both T and $T + c$, the desired temperature field T does not become multivalued or indeterminate. The values of T can be made determinate by appropriate boundary conditions. Conformity to Rule 4 ensures that, if, for example, the boundary temperatures were increased by a constant, all temperatures would increase by exactly that constant.

Another way of looking at Rule 4 is this: When the source term is absent and the neighbor temperatures T_{nb} are all equal, the center temperature T_P must become equal to them. Only a poor discretization equation would not predict $T_P = T_{nb}$ under these circumstances.

3.5 CLOSURE

In this chapter, we have made certain basic decisions about the type of discretization method to be developed in this book. Through a simple example, we have been able to formulate four basic rules, which constitute the underlying guiding principles for all further work. The discussion has been given in terms of temperature T as the dependent variable. This was done simply for conceptual convenience. We shall continue with T in Chapter 4, but switch to the general variable ϕ from Chapter 5 onward. Of course, the four rules developed in this chapter are all applicable to the general variable ϕ.

The convection term in the general differential equation (2.13) requires special formulation. This matter is deferred to Chapter 5. The remaining three terms of Eq. (2.13) are dealt with in Chapter 4 in the framework of heat conduction.

PROBLEMS

3.1 Using the Taylor-series expansion around point P in Fig. 3.2, show that the finite-difference approximation for $d^2 T/dx^2$ is given by

$$\frac{d^2 T}{dx^2} = \frac{2}{(\delta x)_e + (\delta x)_w} \left[\frac{T_E - T_P}{(\delta x)_e} - \frac{T_P - T_W}{(\delta x)_w} \right] .$$

3.2 For the differential equation (3.10), derive a discretization equation by the method of weighted residuals in the following manner: Assume k and S to be constant (for convenience). Let the weighting function W be zero everywhere except between the points W and E in Fig. 3.2. Further, assume that the weighting function is piecewise-linear, with value unity at P and zero at points W and E. Multiply Eq. (3.10) by the weighting function, and integrate over the region from point W to point E. Use a piecewise-linear profile for T. Compare the resulting discretization equation with Eq. (3.12). (Note that the method outlined here, which is a special case of the method of weighted residuals, is known as the Galerkin method.)

3.3 Consider Eq. (3.10) and assume that S is constant, but k depends on x. Further, use a uniform grid spacing in Fig. 3.2, so that $\Delta x = (\delta x)_e = (\delta x)_w$. Derive the discretization equation by writing Eq. (3.10) as

$$k\,\frac{d^2 T}{dx^2} + \frac{dk}{dx}\,\frac{dT}{dx} + S = 0$$

and using the approximations

$$k\,\frac{d^2 T}{dx^2} = \frac{k_P(T_E + T_W - 2T_P)}{(\Delta x)^2}\;,$$

$$\frac{dT}{dx} = \frac{T_E - T_W}{2\,\Delta x}\;,$$

with dk/dx as a given quantity. Noting that dk/dx can be positive or negative, find the conditions for which the coefficient a_E or a_W would become negative, thus violating Rule 2. (Note that the derivation in Section 3.3, which was based on the physical significance of the terms, did not lead to negative coefficients.)

3.4 In an axisymmetrical situation, a steady one-dimensional conduction problem is governed by

$$\frac{1}{r}\,\frac{d}{dr}\left(kr\,\frac{dT}{dr}\right) + S = 0\;,$$

where r is the radial coordinate. Following the procedure in Section 3.3, derive a discretization equation for this situation. (Multiply the differential equation by r, and then integrate with respect to r from r_w to r_e.) Interpret the coefficients in the discretization equation in physical terms.

FOUR

HEAT CONDUCTION

4.1 OBJECTIVES OF THE CHAPTER

In this chapter, we shall begin the task of constructing a numerical method for solving the general differential equation (2.13), which governs the physical processes of interest here. As we have seen, the equation contains four basic terms. Here we shall omit the convection term and concentrate on the remaining three terms. The construction of the method will be completed in Chapter 5, where the treatment of the convection term will be discussed.

Omission of the convection term reduces the situation to a conduction-type problem. Heat conduction provides a convenient starting point for our formulation, because the physical processes are easy to understand and the mathematical complication is minimal.

The objectives of this chapter, however, go far beyond presenting a numerical method for heat conduction alone. First, other physical processes are governed by very similar mathematical equations. Among these are potential flow, mass diffusion, flow through porous media, and some fully developed duct flows. The numerical techniques described in this chapter are directly applicable to all these processes. Electromagnetic field theory, diffusion models of thermal radiation, and lubrication flows are further examples of phenomena governed by conduction-type equations. Although we shall only occasionally make reference to these related processes, it is important to remember that the techniques developed in this chapter are immediately available for application in these different areas.

Second, this chapter accomplishes much of the preparatory work needed for later chapters. The procedure for the solution of the algebraic equations is presented here in a once-and-for-all manner. Later chapters modify the *content* of the algebraic equations, but the same solution technique continues to be applicable. Thus, even for the reader who is exclusively interested in fluid-flow calculation, an understanding of this chapter is essential; much of the material here (and in the next chapter) is an integral part of the fluid-flow calculation scheme to be presented in Chapter 6.

To be able to see the similarities between transfer of momentum and transfer of heat and to regard velocity as, in some ways, analogous to temperature is a great conceptual help. The use of heat conduction as a building block in the fluid-flow calculation scheme reinforces this conceptual unity.

4.2 STEADY ONE–DIMENSIONAL CONDUCTION

4.2-1 The Basic Equations

In the course of presenting the illustrative example in Section 3.3, which was used as a vehicle to explain the four basic rules, we have already derived the discretization equation for steady conduction in one dimension. To review the main ingredients, the governing differential equation is

$$\frac{d}{dx}\left(k\,\frac{dT}{dx}\right) + S = 0 . \tag{4.1}$$

This leads to the discretization equation

$$a_P T_P = a_E T_E + a_W T_W + b , \tag{4.2}$$

where

$$a_E = \frac{k_e}{(\delta x)_e} , \tag{4.3a}$$

$$a_W = \frac{k_w}{(\delta x)_w} , \tag{4.3b}$$

$$a_P = a_E + a_W - S_P\,\Delta x , \tag{4.3c}$$

and

$$b = S_C\,\Delta x . \tag{4.3d}$$

The grid points P, E, and W are shown in Fig. 3.2, where various distances are also indicated. The control-volume faces e and w are placed between the grid point P and its corresponding neighbors. The exact locations of these faces can be considered to be arbitrary. Many practices for their placement are possible, some of which will be discussed in Section 4.6-1. For the time being, we shall simply regard the locations of e and w as *known* in relation to the grid points P, E, and W. The quantities S_C and S_P arise from the source-term linearization of the form

$$S = S_C + S_P T_P . \tag{4.4}$$

As to the profile assumptions, the gradient dT/dx has been evaluated from a piecewise-linear variation of T with x, while for the linearized source term the value T_P is assumed to prevail throughout the control volume. It should, of course, be remembered that other choices of profiles are possible and permissible, as long as the four basic rules are not violated. The policy here is to adopt rather simple profiles within the constraints of these rules and to introduce sophistication only where it is needed.

Many important aspects of the one-dimensional heat-conduction problem still remain to be discussed. It is to these topics that we now turn.

4.2-2 The Grid Spacing

For the grid points shown in Fig. 3.2, it is not necessary that the distances $(\delta x)_e$ and $(\delta x)_w$ be equal. Indeed, the use of nonuniform grid spacing is often desirable, for it enables us to deploy computing power effectively. In general, we shall obtain an accurate solution only when the grid is sufficiently fine. But there is no need to employ a fine grid in regions where the dependent variable T changes rather slowly with x. On the other hand, a fine grid is required where the $T \sim x$ variation is steep.

A misconception seems to prevail that nonuniform grids lead to less accuracy than do uniform grids. There is no sound basis for such an assertion. The grid spacing should be directly linked to the way the dependent variable changes in the calculation domain. Also, there are no universal rules about what maximum (or minimum) ratio the adjacent grid intervals should maintain.

Since the $T \sim x$ distribution is not known before the problem is solved, how can we design an appropriate nonuniform grid? First, one normally has some qualitative expectations about the solution, from which some guidance can be obtained. Second, preliminary coarse-grid solutions can be used to find the pattern of the $T \sim x$ variation; then, a suitable nonuniform grid can be constructed. This is one of the reasons why we insist that our method should give physically meaningful solutions even for coarse grids. An exploratory coarse-grid solution would not be useful if the method gave reasonable solutions only for sufficiently fine grids.

The number of grid points needed for given accuracy and the way they should be distributed in the calculation domain are matters that depend on the nature of the problem to be solved. Exploratory calculations using only a few grid points provide a convenient way of learning about the solution. After all, this is precisely what is commonly done in a laboratory experiment. Preliminary experiments or trial runs are conducted, and the resulting information is used to decide the number and locations of the probes to be installed for the final experiment.

4.2-3 The Interface Conductivity

In Eq. (4.3), the conductivity k_e has been used to represent the value of k pertaining to the control-volume face e; similarly, k_w refers to the interface w. When the conductivity k is a function of x, we shall often know the value of k only at the grid points $W, P, E,$ and so on. We then need a prescription for evaluating the interface conductivity, say k_e, in terms of these grid-point values. The following discussion is, of course, not relevant to situations of uniform conductivity.

Nonuniform conductivity can arise from nonhomogeneity of the material, as in a composite slab. Even in a homogeneous material, the temperature dependence of conductivity can lead to a conductivity variation in response to the temperature distribution. In the treatment of the general differential equation for ϕ, the diffusion coefficient Γ will be handled in the same way as the conductivity k. Significant variations of Γ are frequently encountered, for example, in turbulent flow, where Γ may stand for the turbulent viscosity or turbulent conductivity. Thus, a proper formulation for nonuniform k or Γ is highly desirable.

The most straightforward procedure for obtaining the interface conductivity k_e is to assume a linear variation of k between points P and E. Then,

$$k_e = f_e k_P + (1 - f_e)k_E , \qquad (4.5)$$

where the interpolation factor f_e is a ratio defined in terms of the distances shown in Fig. 4.1:

$$f_e \equiv \frac{(\delta x)_{e+}}{(\delta x)_e} . \qquad (4.6)$$

If the interface e were *midway* between the grid points, f_e would be 0.5, and k_e would be the arithmetic mean of k_P and k_E.

We shall shortly show that this simple-minded approach leads to rather incorrect implications in some cases and cannot accurately handle the abrupt changes of conductivity that may occur in composite materials. Fortunately, a much better alternative of comparable simplicity is available. In developing

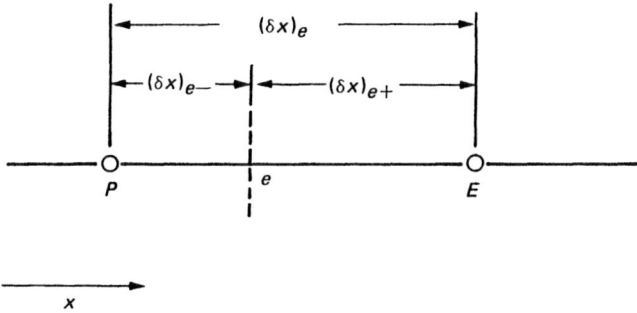

Figure 4.1 Distances associated with the interface e.

this alternative, we recognize that it is not the local value of conductivity at the interface e that concerns us primarily. Our main objective is to obtain a good representation for the heat flux q_e at the interface via

$$q_e = \frac{k_e(T_P - T_E)}{(\delta x)_e} \, , \qquad (4.7)$$

which has, in effect, been used in deriving the discretization equation (4.2). The desired expression for k_e is the one that leads to a "correct" q_e.

Let us consider that the control volume surrounding the grid point P is filled with a material of uniform conductivity k_P, and the one around E with a material of conductivity k_E. For the composite slab between points P and E, a steady one-dimensional analysis (without sources) leads to

$$q_e = \frac{T_P - T_E}{(\delta x)_{e-}/k_P + (\delta x)_{e+}/k_E} \, . \qquad (4.8)$$

Combination of Eqs. (4.6)–(4.8) yields

$$k_e = \left(\frac{1 - f_e}{k_P} + \frac{f_e}{k_E} \right)^{-1} \, . \qquad (4.9)$$

When the interface e is placed *midway* between P and E, we have $f_e = 0.5$; then

$$k_e^{-1} = 0.5(k_P^{-1} + k_E^{-1}) \qquad (4.10a)$$

or
$$k_e = \frac{2k_P k_E}{k_P + k_E} \, . \qquad (4.10b)$$

Equations (4.10) show that k_e is the *harmonic mean* of k_P and k_E, rather than the arithmetic mean which Eq. (4.5) would give when $f_e = 0.5$.

The use of Eq. (4.9) in the coefficient definitions (4.3) leads to the following expression for a_E:

$$a_E = \left[\frac{(\delta x)_{e-}}{k_P} + \frac{(\delta x)_{e+}}{k_E} \right]^{-1}. \qquad (4.11)$$

A similar expression can be written for a_W. Clearly, a_E represents the *conductance* of the material between points P and E.

The effectiveness of this formulation can be quickly seen in the following two limiting cases:

1. Let $k_E \to 0$. Then, from Eq. (4.9),

$$k_e \to 0. \qquad (4.12)$$

This implies that the heat flux at the face of an insulator becomes zero, as it should. The arithmetic-mean formulation, on the other hand, would have given a nonzero flux in this situation.

2. Let $k_P \gg k_E$. Then

$$k_e \to \frac{k_E}{f_e}. \qquad (4.13)$$

This result has two implications; one is easy to understand, and the other is more obscure. Equation (4.13) indicates that the interface conductivity k_e is not at all dependent on k_P. This is to be expected because the high-conductivity material around point P would offer negligible resistance in comparison with the material around E. (The arithmetic-mean formula would have retained the effect of k_P on k_e.) The other implication is that k_e is not equal to k_E, but rather $1/f_e$ times it. A little reflection will show the appropriateness of this. Our purpose is to get a correct value of q_e via Eq. (4.7). The use of Eq. (4.13) yields

$$q_e = \frac{k_E(T_P - T_E)}{(\delta x)_{e+}}. \qquad (4.14)$$

When $k_P \gg k_E$, the temperature T_P will prevail right up to the interface e, and the temperature drop $T_P - T_E$ will actually take place over the distance $(\delta x)_{e+}$. Thus, the correct heat flux will be as given by Eq. (4.14). In other words, the factor f_e in Eq. (4.13) can be seen to compensate for the use of the *nominal* distance $(\delta x)_e$ in Eq. (4.7).

Consideration of these two limiting cases shows that the formulation can handle abrupt changes in the conductivity without requiring an excessively fine grid in the vicinity of the change. This is not only convenient for conduction calculations in composite slabs, but it has other quite fascinating implications. These have been described in Patankar (1978) and will be explained in later chapters.

The recommended interface-conductivity formula (4.9) is based on the steady, no-source, one-dimensional situation in which the conductivity varies in a stepwise fashion from one control volume to the next. Even in situations with nonzero sources or with continuous variation of conductivity, it performs much better than the arithmetic-mean formula. This is demonstrated in Patankar (1978) for some cases for which exact analytical solutions can be found.

4.2-4 Nonlinearity

The discretization equation (4.2) is a linear algebraic equation, and we shall solve the set of such equations by the methods for linear algebraic equations. We shall, however, frequently encounter nonlinear situations even in heat conduction. The conductivity k may depend on T, or the source S may be a nonlinear function of T. Then, the coefficients in the discretization equation will themselves depend on T. We shall handle such situations by iteration. This process involves the following steps:

1. Start with a guess or estimate for the values of T at all grid points.
2. From these guessed T's, calculate tentative values of the coefficients in the discretization equation.
3. Solve the nominally linear set of algebraic equations to get new values of T.
4. With these T's as better guesses, return to step 2 and repeat the process until further repetitions (called iterations) cease to produce any significant changes in the values of T.

This final unchanging state is called the *convergence* of the iterations.[*] The *converged* solution is actually the correct solution of the nonlinear equations, although it is arrived at by the methods for solving linear equations.

It is, however, possible that successive iterations would not ever converge to a solution. The values of T may steadily drift or oscillate with increasing amplitude. This process, which is the opposite of convergence, is called *divergence*. A good numerical method should minimize the possibilities of divergence. As we shall see later, adherence to our four basic rules promotes

---

[*]Sometimes, the term convergence is used for the process by which successive grid refinement brings the numerical solution closer to the exact solution. We shall refer to this aspect as the "accuracy" of the numerical solution, and reserve the word convergence for the convergence of iterations.

convergence; we shall also discuss other strategies for avoiding divergence. At this point, it is sufficient to note that our procedure is not limited to linear problems, and that any nonlinearity can, at least in principle, be handled by the iterative technique just outlined.

4.2-5 Source-Term Linearization

When the source S depends on T, we express the dependence in a linear form given by Eq. (4.4). This is done because (1) our nominally linear framework would allow only a formally linear dependence, and (2) the incorporation of linear dependence is better than treating S as a constant.

When S is a nonlinear function of T, we must *linearize* it, i.e., specify the values of S_C and S_P, which may themselves depend on T. During each iteration cycle, S_C and S_P would then be recalculated from the new values of T. The linearization of S should be a good representation of the $S \sim T$ relationship. Further, the basic rule about nonpositive S_P must be obeyed.

There are many ways of splitting a given expression for S into S_C and $S_P T_P$. Some of these are illustrated by the following examples. The numbers appearing in these examples have no particular significance. The symbol T_P^* is used to denote the guess value or the previous-iteration value of T_P.

Example 1 Given: $S = 5 - 4T$. Some possible linearizations are:

1. $S_C = 5$, $S_P = -4$. This is the most obvious form and is recommended.
2. $S_C = 5 - 4T_P^*$, $S_P = 0$. This is the approach of the lazy person who throws the entire S into S_C and sets S_P equal to zero. This approach, however, is not impracticable and is perhaps the only choice when the expression for S is very complicated.
3. $S_C = 5 + 7T_P^*$, $S_P = -11$. This proposes a steeper $S \sim T$ relationship than the one actually given. The result will be that the convergence of the iterations will slow down. However, if there are other non-linearities in the problem, this slowdown may actually be welcome.

Example 2 Given: $S = 3 + 7T$. Some possible linearizations are:

1. $S_C = 3$, $S_P = 7$. In general this is not acceptable, as it makes S_P positive. If the problem could be solved without iteration, this linearization would give the correct solution, but if iteration is employed for some reason (such as the nonlinearity of other terms), the presence of a positive S_P may cause divergence.
2. $S_C = 3 + 7T_P^*$, $S_P = 0$. This is the practice one should follow when a negative S_P is not naturally forthcoming.
3. $S_C = 3 + 9T_P^*$, $S_P = -2$. This is an artificial creation of a negative S_P. It will, in general, slow down the convergence.

Example 3 Given: $S = 4 - 5T^3$. Some possible linearizations are:

1. $S_C = 4 - 5T_P^{*3}$, $S_P = 0$. This is the lazy-person approach, which fails to take advantage of the known dependence of S on T.
2. $S_C = 4$, $S_P = -5T_P^{*2}$. This looks like the correct linearization, but the given $S \sim T$ curve is steeper than this implies.
3. Recommended method:

$$S = S^* + \left(\frac{dS}{dT}\right)^* (T_P - T_P^*) = 4 - 5T_P^{*3} - 15T_P^{*2}(T_P - T_P^*).$$

Thus,

$$S_C = 4 + 10T_P^{*3}, \qquad S_P = -15T_P^{*2}.$$

This linearization represents the tangent to the $S \sim T$ curve at T_P^*.
4. $S_C = 4 + 20T_P^{*3}$, $S_P = -25T_P^{*2}$. This linearization, which is steeper than the given $S \sim T$ curve, would slow down convergence.

These four possible linearizations are shown in Fig. 4.2 along with the actual $S \sim T$ curve. On such a diagram, straight lines of positive slope would violate basic Rule 3. Among the negative-slope lines, the tangent to the given curve is usually the best choice. Steeper lines are acceptable, but would normally lead to slower convergence. Less steep lines are undesirable, as they fail to incorporate the given rate of fall of S with T.

This discussion of the source-term linearization is adequate for present purposes. Further considerations are given in Chapter 7.

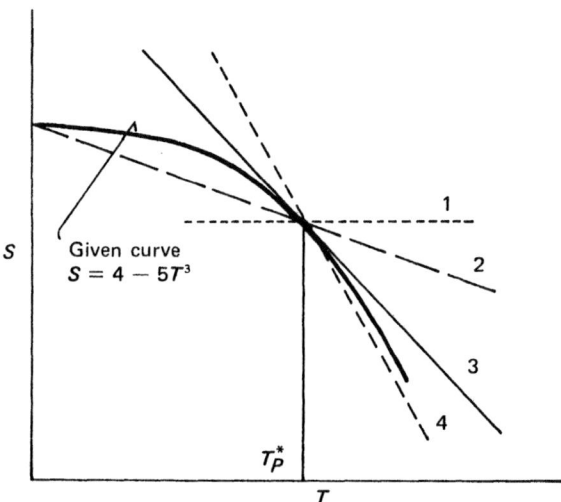

Figure 4.2 The four possible linearizations for Example 3.

4.2-6 Boundary Conditions

Let us consider that, for the one-dimensional problem, the string of grid points shown in Fig. 4.3 is chosen. There is one grid point on each of the two boundaries. The other grid points will be called the *internal* points, around each of which is shown a control volume. A discretization equation like Eq. (4.2) can be written for each such control volume. If Eq. (4.2) is regarded as an equation *for* T_P, we then have the necessary equations for all the unknown temperatures at the internal grid points. Two of these equations, however, involve the boundary grid-point temperatures. It is through the treatment of these boundary temperatures that the given boundary conditions are introduced into the numerical solution scheme.

Since it is not necessary to discuss the two boundary points separately, attention will be focused on the left-hand boundary point B, which is adjacent to the first internal point I as shown in Fig. 4.3. Typically, three kinds of boundary conditions are encountered in heat conduction. These are:

1. Given boundary temperature
2. Given boundary heat flux
3. Boundary heat flux specified via a heat transfer coefficient and the temperature of the surrounding fluid

If the boundary temperature is given (i.e., if the value of T_B is known), no particular difficulty arises, and no additional equations are required. When the boundary temperature is *not* given, we need to construct an additional equation for T_B. This is done by integrating the differential equation over the "half" control volume shown adjacent to the boundary in Fig. 4.3. (This control volume extends only on one side of the grid point B. This is why we refer to it as the half control volume.) An enlarged view of this control volume is given in Fig. 4.4. Integrating Eq. (4.1) over this control volume and noting that the heat flux q stands for $-k\,dT/dx$, we get

$$q_B - q_i + (S_C + S_P T_B)\,\Delta x = 0 , \qquad (4.15)$$

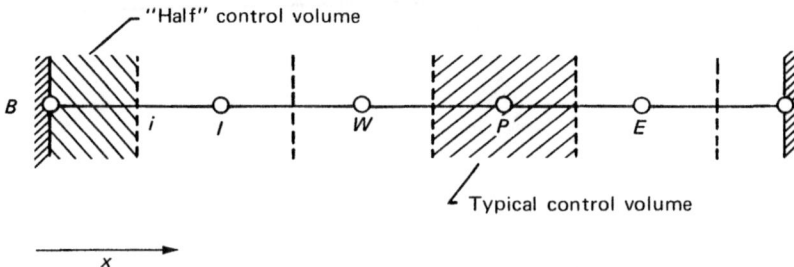

Figure 4.3 Control volumes for the internal and boundary points.

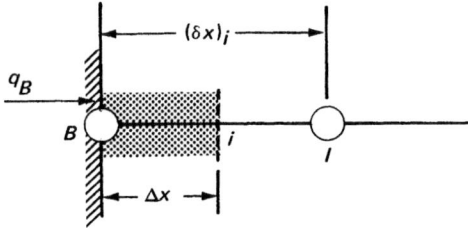

Figure 4.4 Half control volume near the boundary.

where the source term has been linearized in the usual fashion. The interface heat flux q_i can be written along the lines of Eq. (4.7). The result is

$$q_B - \frac{k_i(T_B - T_I)}{(\delta x)_i} + (S_C + S_P T_B) \, \Delta x = 0 \,. \tag{4.16}$$

Further implementation of this equation depends on what is given about the boundary heat flux q_B. If the value of q_B itself is given, the required equation for T_B becomes

$$a_B T_B = a_I T_I + b \,, \tag{4.17}$$

where

$$a_I = \frac{k_i}{(\delta x)_i} \,, \tag{4.18a}$$

$$b = S_C \, \Delta x + q_B \,, \tag{4.18b}$$

$$a_B = a_I - S_P \, \Delta x \,. \tag{4.18c}$$

If the heat flux q_B is specified in terms of a heat transfer coefficient h and a surrounding-fluid temperature T_f such that[*]

$$q_B = h(T_f - T_B) \,, \tag{4.19}$$

then the equation for T_B becomes

$$a_B T_B = a_I T_I + b \,, \tag{4.20}$$

where

$$a_I = \frac{k_i}{(\delta x)_i} \,, \tag{4.21a}$$

[*]It may be recalled that we used the symbol h in Chapter 2 to denote the specific enthalpy. However, no confusion with the heat transfer coefficient h is likely to arise.

$$b = S_C \, \Delta x + h T_f \; , \qquad\qquad (4.21b)$$

$$a_B = a_I - S_P \, \Delta x + h \; . \qquad\qquad (4.21c)$$

In this manner we are able to construct the required number of equations for the unknown temperatures. We shall now describe the method for solving them.

4.2-7 Solution of the Linear Algebraic Equations

The solution of the discretization equations for the one-dimensional situation can be obtained by the standard Gaussian-elimination method. Because of the particularly simple form of the equations, the elimination process turns into a delightfully convenient algorithm. This is sometimes called the Thomas algorithm or the TDMA (*Tri*Diagonal-*M*atrix *A*lgorithm). The designation TDMA refers to the fact that when the matrix of the coefficients of these equations is written, all the nonzero coefficients align themselves along three diagonals of the matrix.

For convenience in presenting the algorithm, it is necessary to use somewhat different nomenclature. Suppose the grid points in Fig. 4.3 were numbered 1, 2, 3, . . . , N, with points 1 and N denoting the boundary points. The discretization equations can be written as

$$a_i T_i = b_i T_{i+1} + c_i T_{i-1} + d_i \; , \qquad\qquad (4.22)$$

for $i = 1, 2, 3, \ldots, N$. Thus, the temperature T_i is related to the neighboring temperatures T_{i+1} and T_{i-1}. To account for the special form of the boundary-point equations, let us set

$$c_1 = 0 \qquad \text{and} \qquad b_N = 0 \; , \qquad\qquad (4.23)$$

so that the temperatures T_0 and T_{N+1} will not have any meaningful role to play. (When the boundary temperatures are given, these boundary-point equations take a rather trivial form. For example, if T_1 is given, we have $a_1 = 1$, $b_1 = 0$, $c_1 = 0$, and $d_1 = $ the given value of T_1.)

These conditions imply that T_1 is known in terms of T_2. The equation for $i = 2$ is a relation between T_1, T_2, and T_3. But, since T_1 can be expressed in terms of T_2, this relation reduces to a relation between T_2 and T_3. In other words, T_2 can be expressed in terms of T_3. This process of substitution can be continued until T_N is formally expressed in terms of T_{N+1}. But, because T_{N+1} has no meaningful existence, we actually obtain the numerical value of T_N at this stage. This enables us to begin the "back-substitution" process in which T_{N-1} is obtained from T_N, T_{N-2} from T_{N-1}, . . . , T_2 from T_3, and T_1 from T_2. This is the essence of the TDMA.

Suppose, in the forward-substitution process, we seek a relation

$$T_i = P_i T_{i+1} + Q_i \tag{4.24}$$

after we have just obtained

$$T_{i-1} = P_{i-1} T_i + Q_{i-1} . \tag{4.25}$$

Substitution of Eq. (4.25) into Eq. (4.22) leads to

$$a_i T_i = b_i T_{i+1} + c_i(P_{i-1} T_i + Q_{i-1}) + d_i , \tag{4.26}$$

which can be rearranged to look like Eq. (4.24). In other words, the coefficients P_i and Q_i then stand for

$$P_i = \frac{b_i}{a_i - c_i P_{i-1}} , \tag{4.27a}$$

$$Q_i = \frac{d_i + c_i Q_{i-1}}{a_i - c_i P_{i-1}} . \tag{4.27b}$$

These are recurrence relations, since they give P_i and Q_i in terms of P_{i-1} and Q_{i-1}. To start the recurrence process, we note that Eq. (4.22) for $i = 1$ is almost of the form (4.24). Thus, the values of P_1 and Q_1 are given by

$$P_1 = \frac{b_1}{a_1} \quad \text{and} \quad Q_1 = \frac{d_1}{a_1} . \tag{4.28}$$

[It is interesting to note that these expressions do follow from Eq. (4.27) after the substitution $c_1 = 0$.]

At the other end of the P_i, Q_i sequence, we note that $b_N = 0$. This leads to $P_N = 0$, and hence from Eq. (4.24) we obtain

$$T_N = Q_N . \tag{4.29}$$

Now we are in a position to start the back substitution via Eq. (4.24).

Summary of the algorithm.

1. Calculate P_1 and Q_1 from Eq. (4.28).
2. Use the recurrence relations (4.27) to obtain P_i and Q_i for $i = 2, 3, \ldots,$ N.
3. Set $T_N = Q_N$.
4. Use Eq. (4.24) for $i = N-1, N-2, \ldots, 3, 2, 1$ to obtain $T_{N-1}, T_{N-2},$ $\ldots, T_3, T_2, T_1$.

The tridiagonal-matrix algorithm is a very powerful and convenient equation solver whenever the algebraic equations can be represented in the form of Eq. (4.22). Unlike general matrix methods, the TDMA requires computer storage and computer time proportional only to N, rather than to N^2 or N^3.

4.3 UNSTEADY ONE–DIMENSIONAL CONDUCTION

4.3-1 The General Discretization Equation

With reference to the general differential equation for ϕ, we have now seen, at least in the one-dimensional context, how to handle the diffusion term and the source term. Here, we turn to the unsteady term and temporarily drop the source term, since nothing new needs to be said about it. Thus, we seek to solve the unsteady one-dimensional heat-conduction equation

$$\rho c \, \frac{\partial T}{\partial t} = \frac{\partial}{\partial x} \left(k \, \frac{\partial T}{\partial x} \right) . \tag{4.30}$$

Further, for convenience, we shall assume ρc to be constant. (In Chapter 2, it was shown how the heat conduction equation could be modified to take account of the variable specific heat c. See Problem 2.2.)

Since time is a one-way coordinate, we obtain the solution by marching in time from a given initial distribution of temperature. Thus, in a typical "time step" the task is this: Given the grid-point values of T at time t, find the values of T at time $t + \Delta t$. The "old" (given) values of T at the grid points will be denoted by T_P^0, T_E^0, T_W^0, and the "new" (unknown) values at time $t + \Delta t$ by T_P^1, T_E^1, T_W^1.

The discretization equation is now derived by integrating Eq. (4.30) over the control volume shown in Fig. 3.2 and over the time interval from t to $t + \Delta t$. Thus,

$$\rho c \int_w^e \int_t^{t+\Delta t} \frac{\partial T}{\partial t} \, dt \, dx = \int_t^{t+\Delta t} \int_w^e \frac{\partial}{\partial x} \left(k \, \frac{\partial T}{\partial x} \right) \, dx \, dt , \tag{4.31}$$

where the order of the integrations is chosen according to the nature of the term. For the representation of the term $\partial T/\partial t$, we shall assume that the grid-point value of T prevails throughout the control volume. Then,

$$\rho c \int_w^e \int_t^{t+\Delta t} \frac{\partial T}{\partial t} \, dt \, dx = \rho c \, \Delta x \, (T_P^1 - T_P^0) . \tag{4.32}$$

Following our steady-state practice for $k \, \partial T/\partial x$, we obtain

$$\rho c \, \Delta x \, (T_P^1 - T_P^0) = \int_t^{t+\Delta t} \left[\frac{k_e(T_E - T_P)}{(\delta x)_e} - \frac{k_w(T_P - T_W)}{(\delta x)_w} \right] dt \, . \quad (4.33)$$

It is at this point that we need an assumption about how T_P, T_E, and T_W vary with time from t to $t + \Delta t$. Many assumptions are possible, and some of them can be generalized by proposing

$$\int_t^{t+\Delta t} T_P \, dt = [fT_P^1 + (1 - f)T_P^0] \, \Delta t \, , \quad (4.34)$$

where f is a weighting factor between 0 and 1. Using similar formulas for the integrals of T_E and T_W, we derive from Eq. (4.33)

$$\rho c \, \frac{\Delta x}{\Delta t} \, (T_P^1 - T_P^0) = f \left[\frac{k_e(T_E^1 - T_P^1)}{(\delta x)_e} - \frac{k_w(T_P^1 - T_W^1)}{(\delta x)_w} \right]$$
$$+ (1 - f) \left[\frac{k_e(T_E^0 - T_P^0)}{(\delta x)_e} - \frac{k_w(T_P^0 - T_W^0)}{(\delta x)_w} \right] \, . \quad (4.35)$$

While rearranging this, we shall drop the superscript 1, and remember that T_P, T_E, T_W henceforth stand for the new values of T at time $t + \Delta t$. The result is

$$a_P T_P = a_E \, [fT_E + (1 - f)T_E^0] + a_W \, [fT_W + (1 - f)T_W^0]$$
$$+ [a_P^0 - (1 - f)a_E - (1 - f)a_W] \, T_P^0 \, , \quad (4.36)$$

where

$$a_E = \frac{k_e}{(\delta x)_e} \, , \quad (4.37a)$$

$$a_W = \frac{k_w}{(\delta x)_w} \, , \quad (4.37b)$$

$$a_P^0 = \frac{\rho c \, \Delta x}{\Delta t} \, , \quad (4.37c)$$

$$a_P = fa_E + fa_W + a_P^0 \, . \quad (4.37d)$$

4.3-2 Explicit, Crank-Nicolson, and Fully Implicit Schemes

For certain specific values of the weighting factor f, the discretization equation reduces to one of the well-known schemes for parabolic differential equations. In particular, $f = 0$ leads to the explicit scheme, $f = 0.5$ to the Crank-Nicolson scheme, and $f = 1$ to the fully implicit scheme. We shall briefly discuss these schemes and finally indicate the fully implicit scheme as our preference.

The different values of f can be interpreted in terms of the $T_P \sim t$ variations shown in Fig. 4.5. The explicit scheme essentially assumes that the old value T_P^0 prevails throughout the entire time step except at time $t + \Delta t$. The fully implicit scheme postulates that, at time t, T_P suddenly drops from T_P^0 to T_P^1 and then stays at T_P^1 over the whole of the time step; thus the temperature during the time step is characterized by T_P^1, the new value. The Crank-Nicolson scheme assumes a linear variation of T_P. At first sight, the linear variation would appear more sensible than the two other alternatives. Why then would we prefer the fully implicit scheme? The answer will emerge very shortly.

For the explicit scheme ($f = 0$), Eq. (4.36) becomes

$$a_P T_P = a_E T_E^0 + a_W T_W^0 + (a_P^0 - a_E - a_W) T_P^0 . \qquad (4.38)$$

This means that T_P is not related to other unknowns such as T_E or T_W, but is *explicitly* obtainable in terms of the known temperatures T_P^0, T_E^0, T_W^0. This is why the scheme is called explicit. Any scheme with $f \neq 0$ would be implicit;

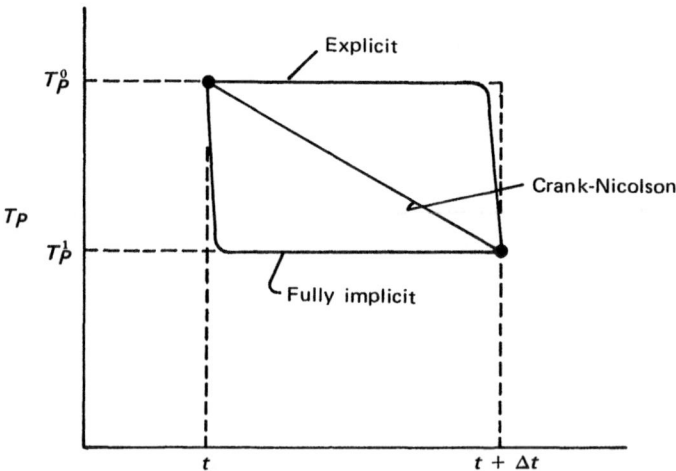

Figure 4.5 Variation of temperature with time for three different schemes.

that is, T_P would be linked to the unknowns T_E and T_W, and the solution of a set of simultaneous equations would be necessary. The convenience of the explicit scheme in this regard is, however, offset by a serious limitation. If we remember the basic rule about positive coefficients (Rule 2) and examine Eq. (4.38), we note that the coefficient of T_P^0 can become negative. (We consider T_P^0 as a neighbor of T_P in the time direction.) Indeed, for this coefficient to be positive, the time step Δt would have to be small enough so that a_P^0 exceeds $a_E + a_W$. For uniform conductivity and $\Delta x = (\delta x)_e = (\delta x)_w$, this condition can be expressed as

$$\Delta t < \frac{\rho c (\Delta x)^2}{2k} . \tag{4.39}$$

If this condition is violated, physically unrealistic results could emerge, because the negative coefficient implies that a higher T_P^0 results in a lower T_P. Equation (4.39) is the well-known stability criterion for the explicit scheme. It is interesting to note that we have been able to derive this from physical arguments based on one of our basic rules. The troublesome feature about condition (4.39) is that, as we reduce Δx to improve the spatial accuracy, we are forced to use a much smaller Δt.

The Crank-Nicolson scheme is usually described as unconditionally stable. An inexperienced user often interprets this to imply that a physically realistic solution will result no matter how large the time step, and such a user is, therefore, surprised to encounter oscillatory solutions. The "stability" in a mathematical sense simply ensures that these oscillations will eventually die out, but it does not guarantee physically plausible solutions. Some examples of unrealistic solutions given by the Crank-Nicolson scheme can be found in Patankar and Baliga (1978).

In our framework, this behavior is easy to explain. For $f = 0.5$, the coefficient of T_P^0 in Eq. (4.36) becomes $a_P^0 - (a_E + a_W)/2$. For uniform conductivity and uniform grid spacing, this coefficient can be seen to be $\rho c \, \Delta x/\Delta t - k/\Delta x$. Again, whenever the time step is not sufficiently small, this coefficient could become negative, with its potential for physically unrealistic results. The seemingly reasonable *linear* profile in Fig. 4.5 is a good representation of the temperature-time relationship for only small time intervals. Over a larger interval, the intrinsically exponential decay of temperature is akin to a steep drop in the beginning, followed by a flat tail. The assumptions made in the fully implicit scheme are thus closer to reality than the linear profile used in the Crank-Nicolson scheme, especially for large time steps.

If we require that the coefficient of T_P^0 in Eq. (4.36) must never become negative, the only constant value of f that ensures this is 1. (Of course, it is not meaningful for f to be greater than 1.) Thus, the fully implicit scheme ($f = 1$) satisfies our requirements of simplicity and physically satisfactory behavior. It is for this reason that we shall adopt the fully implicit scheme in this book.

It must be admitted that for small time steps the fully implicit scheme is not as accurate as the Crank-Nicolson scheme. Again, the reason can be seen from Fig. 4.5; the temperature-time curve *is* nearly linear for small time intervals. It is tempting to seek a scheme that combines the advantages of both schemes and shares the disadvantages of neither. Indeed, this has been done, and the result, called the *exponential* scheme, has been described by Patankar and Baliga (1978). That scheme, however, is somewhat complicated, and its inclusion in this book, in which many other themes are to be presented, would have made the treatment quite intricate.

4.3-3 The Fully Implicit Discretization Equation

Here we record the fully implicit form of Eq. (4.36). In doing so, we shall introduce the linearized source term, which we had temporarily dropped. The result is

$$a_P T_P = a_E T_E + a_W T_W + b ,$$ (4.40)

where

$$a_E = \frac{k_e}{(\delta x)_e} ,$$ (4.41a)

$$a_W = \frac{k_w}{(\delta x)_w} ,$$ (4.41b)

$$a_P^0 = \frac{\rho c\,\Delta x}{\Delta t} ,$$ (4.41c)

$$b = S_C\,\Delta x + a_P^0 T_P^0 ,$$ (4.41d)

$$a_P = a_E + a_W + a_P^0 - S_P\,\Delta x .$$ (4.41e)

It can be seen that, as $\Delta t \to \infty$, this equation reduces to our steady-state discretization equation.

The main principle of the fully implicit scheme is that the new value T_P prevails over the entire time step. Thus, if the conductivity k_P depended on temperature, it should be iteratively recalculated from T_P, exactly as in our steady-state procedure. Other aspects of the steady-state procedure, such as boundary conditions, source-term linearization, and the TDMA, are also equally applicable to the unsteady situation.

Our detailed consideration of the one-dimensional problem has now set the stage for extension to two and three dimensions. The extension is surprisingly easy.

4.4 TWO- AND THREE-DIMENSIONAL SITUATIONS

4.4-1 Discretization Equation for Two Dimensions

A portion of a two-dimensional grid is shown in Fig. 4.6. For the grid point P, points E and W are its x-direction neighbors, while N and S (denoting north and south) are the y-direction neighbors. The control volume around P is shown by dashed lines. Its thickness in the z direction is assumed to be unity. The nomenclature introduced in Fig. 3.2 for distances Δx, $(\delta x)_e$, etc. is to be extended to two dimensions here. The question of the actual location of the control-volume faces in relation to the grid points is still left open. Locating them exactly *midway* between the neighboring grid points is an obvious possibility, but other practices can also be employed, some of which will be discussed in Section 4.6-1. Here we shall derive discretization equations that will be applicable to any such practice.

We have seen how to calculate the heat flux q_e at the control-volume face between P and E. We shall assume that q_e, thus obtained, prevails over the entire face of area $\Delta y \times 1$. Heat flow rates through the other faces can be obtained in a similar fashion. In this manner, the differential equation

$$\rho c \frac{\partial T}{\partial t} = \frac{\partial}{\partial x}\left(k \frac{\partial T}{\partial x}\right) + \frac{\partial}{\partial y}\left(k \frac{\partial T}{\partial y}\right) + S \qquad (4.42)$$

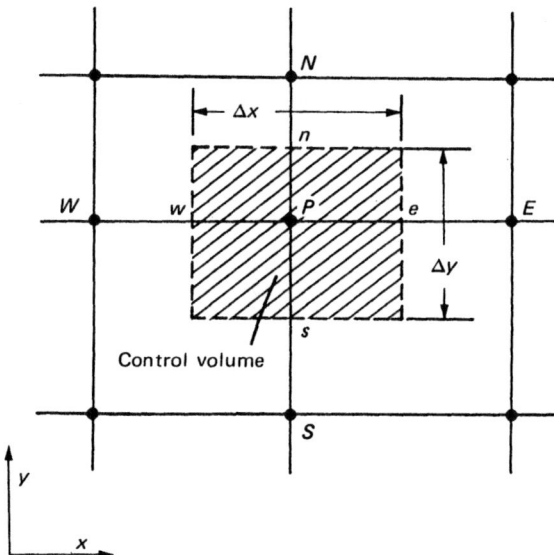

Figure 4.6 Control volume for the two-dimensional situation.

can be instantly turned into the discretization equation

$$a_P T_P = a_E T_E + a_W T_W + a_N T_N + a_S T_S + b , \qquad (4.43)$$

where

$$a_E = \frac{k_e \, \Delta y}{(\delta x)_e} , \qquad (4.44a)$$

$$a_W = \frac{k_w \, \Delta y}{(\delta x)_w} , \qquad (4.44b)$$

$$a_N = \frac{k_n \, \Delta x}{(\delta y)_n} , \qquad (4.44c)$$

$$a_S = \frac{k_s \, \Delta x}{(\delta y)_s} , \qquad (4.44d)$$

$$a_P^0 = \frac{\rho c \, \Delta x \, \Delta y}{\Delta t} , \qquad (4.44e)$$

$$b = S_C \, \Delta x \, \Delta y + a_P^0 T_P^0 , \qquad (4.44f)$$

$$a_P = a_E + a_W + a_N + a_S + a_P^0 - S_P \, \Delta x \, \Delta y . \qquad (4.44g)$$

The product $\Delta x \, \Delta y$ is the volume of the control volume.

4.4-2 Discretization Equation for Three Dimensions

Finally, we add two more neighbors T and B (top and bottom) for the z direction to complete the three-dimensional configuration. The discretization equation can easily be seen to be

$$a_P T_P = a_E T_E + a_W T_W + a_N T_N + a_S T_S + a_T T_T + a_B T_B + b , \quad (4.45)$$

where

$$a_E = \frac{k_e \, \Delta y \, \Delta z}{(\delta x)_e} , \qquad (4.46a)$$

$$a_W = \frac{k_w \, \Delta y \, \Delta z}{(\delta x)_w} , \qquad (4.46b)$$

$$a_N = \frac{k_n \, \Delta z \, \Delta x}{(\delta y)_n} , \qquad (4.46c)$$

$$a_S = \frac{k_s \, \Delta z \, \Delta x}{(\delta y)_s} \quad , \tag{4.46d}$$

$$a_T = \frac{k_t \, \Delta x \, \Delta y}{(\delta z)_t} \quad , \tag{4.46e}$$

$$a_B = \frac{k_b \, \Delta x \, \Delta y}{(\delta z)_b} \quad , \tag{4.46f}$$

$$a_P^0 = \frac{\rho c \, \Delta x \, \Delta y \, \Delta z}{\Delta t} \quad , \tag{4.46g}$$

$$b = S_C \, \Delta x \, \Delta y \, \Delta z + a_P^0 T_P^0 \quad , \tag{4.46h}$$

$$a_P = a_E + a_W + a_N + a_S + a_T + a_B + a_P^0 - S_P \, \Delta x \, \Delta y \, \Delta z \quad . \tag{4.46i}$$

At this point, it is interesting to examine the physical significance of the various coefficients in the discretization equation. The neighbor coefficients a_E, a_W, a_N, ..., a_B represent the conductance between the point P and the corresponding neighbor. The term $a_P^0 T_P^0$ is the internal energy (divided by Δt) contained in the control volume at time t. The constant term b consists of this internal energy and the rate of heat generation in the control volume resulting from S_C. The center-point coefficient a_P is the sum of all neighbor coefficients (including a_P^0, which is the coefficient of the "time neighbor" T_P^0) and contains a contribution from the linearized source term.

4.4-3 Solution of the Algebraic Equations

It should be noted that, while constructing the discretization equations, we cast them into a linear form but did not assume that a particular method would be used for their solution. Therefore, any suitable solution method can be employed at this stage. It is useful to consider the derivation of the equations and their solution as two distinct operations, and there is no need for the choices in one to influence the other. In a computer program, the two operations can be conveniently performed in separate sections, and either section can be independently modified when desired.

So far, we have obtained the multidimensional discretization equations by a straightforward extension of the one-dimensional situation. One procedure that cannot so easily be extended to multiple dimensions is the tridiagonal-matrix algorithm (TDMA). Direct methods (i.e., those requiring no iteration) for solving the algebraic equations arising in two- or three-dimensional problems are much more complicated and require rather large amounts of computer storage and time. For a linear problem, which requires the solution of the algebraic equations *only once*, a direct method may be acceptable; but in nonlinear problems, since the equations have to be solved repeatedly with updated coefficients, the use of a direct method is usually not economical. We shall, therefore, exclude direct methods from further consideration, except to say that a computer program for the direct solution of discretization equations in two dimensions has been published by King (1976).

The alternative, then, is iterative methods for the solution of algebraic equations. These start from a guessed field of T (the dependent variable) and use the algebraic equations in some manner to obtain an improved field. Successive repetitions of the algorithm finally lead to a solution that is sufficiently close to the correct solution of the algebraic equations. Iterative methods usually require very small additional storage in the computer, and they are especially attractive for handling nonlinearities. In a nonlinear problem, it is not necessary or wise to take the solution of the algebraic equations to final convergence for a fixed set of coefficient values. With a given set of these values, a few iterations of the equation-solving algorithm are sufficient before the updating of the coefficients is performed. It seems that, in general, there should be a certain balance between the effort required to calculate the coefficients and that spent on solving the equations. Once the coefficients are calculated, we must perform sufficient iterations of the equation solver to extract substantial benefit from the coefficients, but it is unwise to spend an excessive amount of effort on solving equations that are based on only tentative coefficients.

There are many iterative methods for solving algebraic equations. We shall describe only two methods; the first will set the background, and the second is recommended for use.

The Gauss-Seidel point-by-point method   The simplest of all iterative methods is the Gauss-Seidel method in which the values of the variable are calculated by visiting each grid point in a certain order. Only one set of T's is held in computer storage. In the beginning, these represent the initial guess or values from the previous iteration. As each grid point is visited, the corresponding value of T in the computer storage is altered as follows: If the discretization equation is written as

$$a_P T_P = \Sigma \, a_{nb} T_{nb} + b \, , \qquad (4.47)$$

where the subscript nb denotes a neighbor point, then T_P at the visited grid point is calculated from

$$T_P = \frac{\Sigma \, a_{nb} T_{nb}^* + b}{a_P} \, , \qquad (4.48)$$

where T_{nb}^* stands for the neighbor-point value present in the computer storage. For neighbors that have already been visited during the current iteration, T_{nb}^* is the freshly calculated value; for yet-to-be-visited neighbors, T_{nb}^* is the value from the previous iteration. In any case, T_{nb}^* is the latest available value for the neighbor-point temperature. When all grid points have been visited in this manner, one iteration of the Gauss-Seidel method is complete.

To illustrate the method, we shall consider two very simple examples.

Equations:

$$T_1 = 0.4T_2 + 0.2 , \qquad (4.49a)$$

$$T_2 = T_1 + 1 . \qquad (4.49b)$$

Solution:

Iteration no.	0	1	2	3	4	5	$\cdots$	∞
T_1	0	0.2	0.68	0.872	0.949	0.980	$\cdots$	1.0
T_2	0	1.2	1.68	1.872	1.949	1.980	$\cdots$	2.0

It can be seen that, starting with an arbitrary guess, we have been able to approach the correct solution of the equations. An interesting feature of iterative methods is that the accuracy of the calculations may not be very high in the intermediate stages. Approximate calculations, and even errors, tend to be wiped out, since the intermediate values are used simply as guesses for the next iteration. We can gain further insight from the following example.

Equations:

$$T_1 = T_2 - 1 , \qquad (4.50a)$$

$$T_2 = 2.5T_1 - 0.5 . \qquad (4.50b)$$

Solution:

Iteration no.	0	1	2	3	4
T_1	0	−1	−4	−11.5	−30.25
T_2	0	−3	−10.5	−29.25	−76.13

This does not look very hopeful. Here the iteration process has *diverged*. What is more surprising is that Eqs. (4.50) are simply rearranged versions of Eqs. (4.49), for which we did get convergence.

We thus conclude that the Gauss-Seidel method does not always converge. Indeed, a criterion has been formulated by Scarborough (1958) that, when satisfied, guarantees the convergence of the Gauss-Seidel method. We shall state it without proof and discuss its implications.

The Scarborough criterion. A *sufficient* condition for the convergence of the Gauss-Seidel method is

$$\frac{\Sigma|a_{nb}|}{|a_P|} \begin{cases} \leqslant 1 & \text{for all equations} \qquad\qquad (4.51a) \\ < 1 & \text{for at least one equation .} \qquad (4.51b) \end{cases}$$

Comments. (1) The criterion is a sufficient condition, not a *necessary* one. This means that we can, at times, violate the criterion and still obtain convergence. (2) Although we shall not advocate the use of the Gauss-Seidel method, it seems desirable that our discretization equations should satisfy the Scarborough criterion so that convergence is assured by at least one iterative method. (3) Some of our basic rules, which have been motivated by physical considerations, can now be seen to fulfill the demands of the Scarborough criterion. For example, the presence of a negative S_P leads to $\Sigma a_{nb}/a_P < 1$. Our requirement of positive coefficients can also be viewed in this light. If some of the coefficients were negative, then a_P (which often equals Σa_{nb}) could have a magnitude less than $\Sigma|a_{nb}|$ (since $\Sigma a_{nb} < \Sigma|a_{nb}|$), thus leading to a violation of the criterion. (4) When a_P equals Σa_{nb} and all the coefficients are positive, we obtain, for all equations, $\Sigma|a_{nb}|/|a_P| = 1$. Where, then, is the equation at least for which $\Sigma|a_{nb}|/|a_P|$ would become less than unity? The answer lies in the boundary conditions. For the problem to have a determinate solution, the temperature must be specified for at least one boundary point. The discretization equation in which this point appears as one of the neighbors does imply $\Sigma|a_{nb}|/|a_P| < 1$. This is so because $\Sigma|a_{nb}|$ should be calculated, for the purpose of using the Scarborough criterion, as the sum of the coefficients of only the *unknown* neighbors; a_P, on the other hand, is the sum of all neighbor coefficients including the boundary-point coefficient.

A major disadvantage of the otherwise attractive Gauss-Seidel method is that its convergence is too slow, especially when a large number of grid points are involved. The reason for the slowness is easy to understand; the method transmits the boundary-condition information at a rate of one grid interval per iteration.

A line-by-line method A convenient combination of the direct method (TDMA) for one-dimensional situations and the Gauss-Seidel method can now be formed. We shall choose a grid line (say, in the y direction), assume that the T's along the neighboring lines (i.e., the x- and z-direction neighbors of the points on the chosen line) are known from their "latest" values, and solve for the T's along the chosen line by the TDMA. We shall follow this procedure for all the lines in one direction and repeat the procedure, if desired, for the lines in the other direction(s). Although the method is equally applicable to two or three dimensions, we shall, for convenience, conduct the following discussion for two-dimensional situations.

Discussion. (1) The line-by-line scheme can be easily visualized with reference to Fig. 4.7. The discretization equations for the grid points along a chosen line are considered. They contain the temperatures at the grid points (shown by crosses) along the two neighboring lines. If these temperatures are substituted from their latest values, the equations for the grid points (shown by dots) along the chosen line would look like one-dimensional equations and could be solved by the TDMA. This procedure is carried out for all the lines in the y direction and may be followed by a similar treatment for the x direction. (2) The convergence of the line-by-line method is faster, because the boundary-condition information from the ends of the line is transmitted *at once* to the interior of the domain, no matter how many grid points lie along the line. The rate of transmission of information in the other direction is similar to that of the point-by-point method. (3) By alternating the directions in which the TDMA traverse is employed, we can quickly bring the information from all boundaries to the interior. (4) Often the geometry and other properties of the situation result in, for example, the y-direction coefficients being much larger than the x-direction coefficients (see Fig. 4.8). In such a case, especially fast convergence is obtained when the TDMA traverse is employed in the y direction (the direction of larger coefficients). This is because the guess values substituted for the temperatures along the

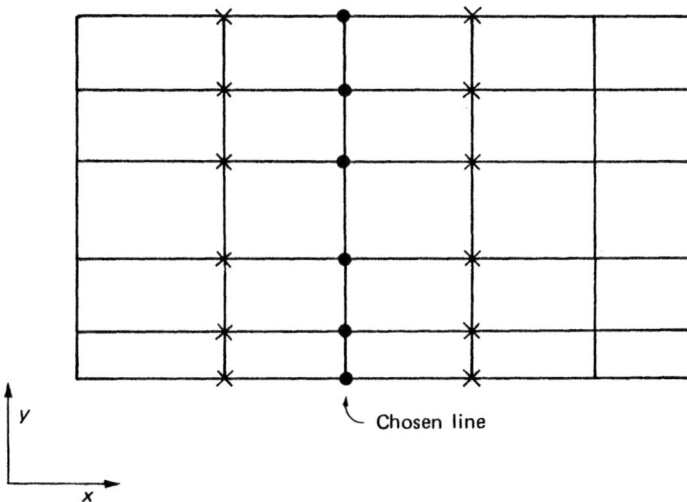

Figure 4.7 Representation of the line-by-line method.

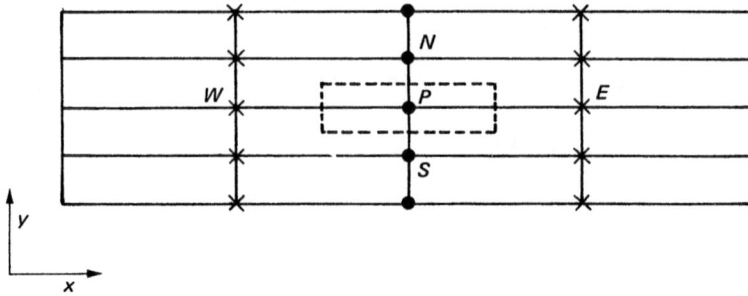

Figure 4.8 Situation in which the y-direction coefficients are much larger than the x-direction coefficients.

neighboring lines have insignificant influence on the discretization equations. (5) In addition to the traverse direction, the sweep direction (i.e., the sequence in which lines are chosen) is also important in some cases. For the boundary conditions shown in Fig. 4.9, a left-to-right sweep (i.e., choosing the left boundary of the domain as the first line and then moving successively to the lines to the right) would transmit the known temperature on the left boundary into the domain; on the other hand, since no temperatures are given on the right boundary, a right-to-left sweep would bring no such useful information. (The same consideration applies to the sequence in which points are visited in a point-by-point scheme.) The sweep direction is especially important when convection is present. Quite clearly, a sweep from upstream to downstream would produce much faster convergence than a sweep against the stream.

Other iterative methods A commonly used line-by-line method known as ADI (*A*lternating-*D*irection *I*mplicit) was introduced by Peaceman and Rachford (1955). Another iterative technique for solving multidimensional discretization equations is the *S*trongly *I*mplicit *P*rocedure (SIP) described by Stone (1968). A detailed study of these methods is left to the interested reader.

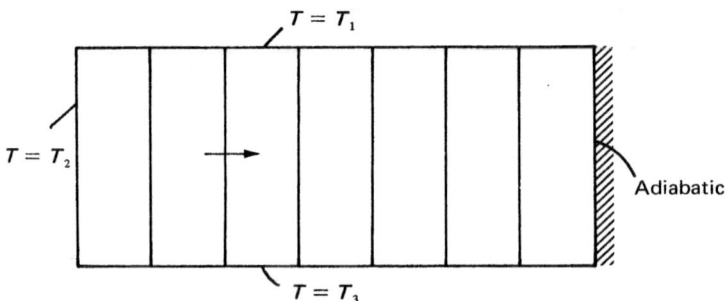

Figure 4.9 Boundary conditions that make a left-to-right sweep more advantageous.

4.5 OVERRELAXATION
AND UNDERRELAXATION

In the iterative solution of the algebraic equations or in the overall iterative scheme employed for handling nonlinearity, it is often desirable to speed up or to slow down the changes, from iteration to iteration, in the values of the dependent variable. This process is called *overrelaxation* or *underrelaxation*, depending on whether the variable changes are accelerated or slowed down. Overrelaxation is often used in conjunction with the Gauss-Seidel method, the resulting scheme being known as *Successive Over-Relaxation* (SOR). With the line-by-line method, the use of overrelaxation is less common. Underrelaxation is a very useful device for nonlinear problems. It is often employed to avoid divergence in the iterative solution of strongly nonlinear equations.

There are many ways of introducing overrelaxation or underrelaxation. Some practices will be described here. We shall work with the general discretization equation of the form

$$a_P T_P = \Sigma \, a_{nb} T_{nb} + b \, . \tag{4.52}$$

Further, T_P^* will be taken as the value of T_P from the previous iteration.

Use of a relaxation factor. Equation (4.52) can be written as

$$T_P = \frac{\Sigma \, a_{nb} T_{nb} + b}{a_P} \, . \tag{4.53}$$

If we add T_P^* to the right-hand side and subtract it, we have

$$T_P = T_P^* + \left(\frac{\Sigma \, a_{nb} T_{nb} + b}{a_P} - T_P^* \right) \, , \tag{4.54}$$

where the contents of the parentheses represent the change in T_P produced by the current iteration. This change can be modified by the introduction of a relaxation factor α, so that

$$T_P = T_P^* + \alpha \left(\frac{\Sigma \, a_{nb} T_{nb} + b}{a_P} - T_P^* \right) \, , \tag{4.55a}$$

or $$\frac{a_P}{\alpha} \, T_P = \Sigma \, a_{nb} T_{nb} + b + (1 - \alpha) \frac{a_P}{\alpha} \, T_P^* \, . \tag{4.55b}$$

At first, it should be noted that, when the iterations converge, that is, T_P becomes equal to T_P^*, Eq. (4.55a) implies that the converged values of T do satisfy the original equation (4.52). Any relaxation scheme, of course, must possess this property; the final converged solution, although obtained through the use of arbitrary relaxation factors or similar devices, must still satisfy the original discretization equation.

When the relaxation factor α in Eq. (4.55) is between 0 and 1, its effect is underrelaxation; that is, the values of T_P stay closer to T_P^*. For a very small value of α, the changes in T_P become very slow. When α is greater than 1, overrelaxation is produced.

There are no general rules for choosing the best value of α. The optimum value depends upon a number of factors, such as the nature of the problem, the number of grid points, the grid spacing, and the iterative procedure used. Usually, a suitable value of α can be found by experience and from exploratory computations for the given problem.

There is no need to maintain the same value of α during the entire computation. The value can be changed from iteration to iteration. Indeed, it is permissible, though not very convenient, to choose a different value of α for each grid point.

Relaxation through inertia. Another technique of overrelaxation or underrelaxation is to replace the discretization equation (4.52) with

$$(a_P + i)T_P = \Sigma \, a_{nb} T_{nb} + b + iT_P^* , \qquad (4.56)$$

where i is the so-called *inertia*. For positive values of i, Eq. (4.56) has the effect of underrelaxation, while negative values of i produce overrelaxation.

Again, there are no general rules for finding the optimum value of the inertia i; it must be determined from experience with a particular problem. From Eq. (4.56), we can deduce that i should be comparable to a_P, and the greater the magnitude of i the stronger will be the effect of the relaxation.

Sometimes, the solution of a steady-state problem is obtained through the use of the discretization equations for a corresponding unsteady situation. Then the "time steps" become the same as iterations, and the "old" value T_P^0 simply represents the previous-iteration value T_P^*. In this sense, the term $a_P^0 T_P^0$ in Eq. (4.46h) acts in the same way as the term iT_P^* in Eq. (4.56). Thus, the inertia i is analogous to the coefficient a_P^0 in the unsteady formulation. This analogy suggests one way of deciding on a reasonable value of i. On the other hand, the practice of solving a steady-state problem via the unsteady formulation can now be recognized as simply a particular kind of underrelaxation procedure. The smaller the time step chosen, the stronger is the resulting underrelaxation. Incidentally, a negative value of the time step Δt would produce overrelaxation.

4.6 SOME GEOMETRIC CONSIDERATIONS

4.6-1 Location of the Control-Volume Faces

So far, no specific information has been provided as to where the control-volume faces are to be located in relation to the grid points. The derivation of the discretization equation has been conducted in general terms so that it will be applicable to any particular way of locating the control-volume faces. Among the many possible practices, we shall look at two different alternatives and discuss their relative merits. The two practices will be called Practice A and Practice B. For convenience, the description will refer to a two-dimensional situation, although the concepts involved are applicable to one- and three-dimensional situations as well.

Practice A: faces located midway between the grid points. The most obvious way of constructing the control volumes is to place their faces *midway* between neighboring grid points. This is shown in Fig. 4.10, where the dashed lines indicate the control-volume faces. The grid is deliberately drawn to be highly nonuniform; one consequence is that a typical grid point *P*, it can be observed, does not lie at the geometric center of the control volume that surrounds it.

Practice B: grid points placed at the centers of the control volumes. Another practice, illustrated in Fig. 4.11, is to draw the control-volume boundaries first and then place a grid point at the geometric center of each control volume. In this scheme, when the control-volume sizes are nonuniform, their faces do not lie midway between the grid points.

Discussion. (1) It should be noted that for uniform grids (or uniform control-volume sizes) the two practices become identical. Therefore, a comparison of the two practices is meaningful only in the context of nonuniform grid spacing. (2) The "midway" faces in Practice A do provide greater accuracy in calculating the heat flux across the face. As noted in Section 3.4, the slope of the piecewise-linear temperature profile happens to be the same as the slope of any parabolic profile evaluated midway between the grid points. Thus, even though a linear profile is used, the results effectively correspond to a less crude parabolic profile. (3) On the other hand, the fact that the grid point *P* in Fig. 4.10 may not be at the geometric center of the

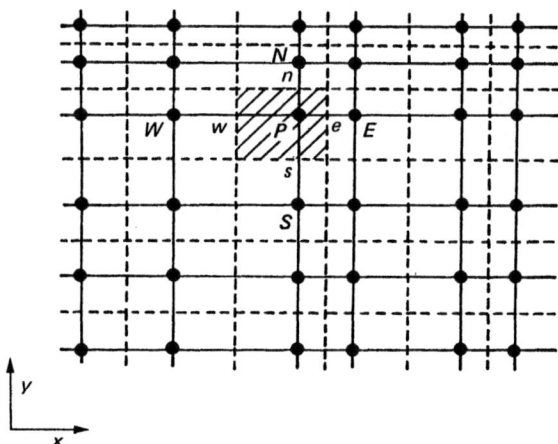

Figure 4.10 Locations of the control-volume faces for Practice A.

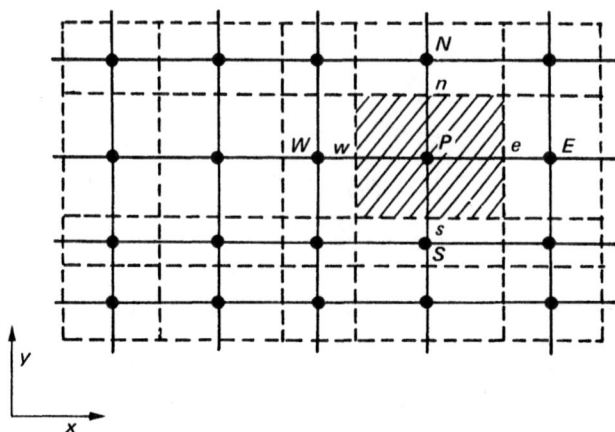

Figure 4.11 Locations of the control-volume faces for Practice B.

control volume represents a disadvantage. The temperature T_P then cannot be regarded as a good representative value for the control volume in the calculation of the source term, the conductivity, and similar quantities. Further, even in the calculation of the heat fluxes at the control-volume faces, Practice A is not free from objections. The point e in Fig. 4.10, for example, is not at the center of the control-volume face on which it lies. Then, to assume that the heat flux at e prevails over the entire face entails some inaccuracy. (4) Practice B does not have these shortcomings, since the point P lies, by definition, at the center of the control volume, and points such as e lie at the center of their respective faces (see Fig. 4.11). The faces, however, do not lie midway between the grid points, and therefore, unlike Practice A, Practice B does not benefit from the fortuitous property of the parabola. (5) Perhaps the decisive advantage of Practice B is the convenience it offers. Since the control volume turns out to be the basic unit of the discretization method developed so far, it is more convenient to draw the control-volume boundaries first and let the grid-point locations follow as a consequence. For a composite solid, for example, we can locate the control-volume faces where the discontinuity in the material properties occurs (see Fig. 4.12). Similarly, discontinuities in boundary conditions can be conveniently handled. If a part of the boundary is adiabatic and the rest isothermal, the control volume can be designed so as to avoid the presence of the discontinuity *within* a control-volume face; this is shown in Fig. 4.12. In Practice A, it is much more difficult to arrange that the control-volume faces fall at the desired locations, because one must first specify the positions of the grid points. (6) The design of the control volumes near the boundaries of the calculation domain requires additional consideration. As shown in Fig. 4.13, Practice A leads to the "half" control volumes (introduced in Section 4.2-6) around the boundary grid points. In Practice B, it is convenient to

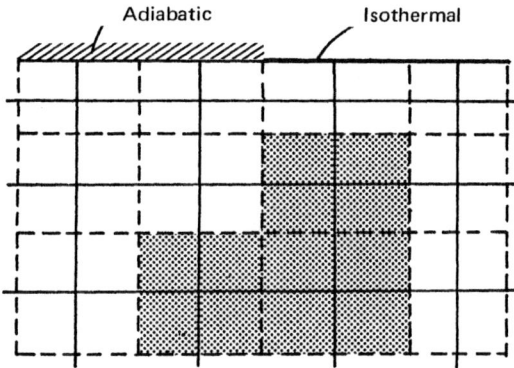

Figure 4.12 Treatment of composite material and discontinuous boundary conditions in Practice B.

completely fill the calculation domain with regular control volumes and to place the boundary grid points on the faces of the near-boundary control volumes. This arrangement is shown in Fig. 4.14. A typical boundary face i is not located *between* the boundary point B and the internal point I, but actually passes through the boundary point. If a control volume of zero thickness is imagined around point B, the location of the face i in relation to the grid points B and I can be seen to conform to the general pattern of Practice B. With such an arrangement, there is no need for the special discretization equation for the near-boundary control volume; the available boundary-condition data, such as given temperature or heat flux, can be directly used at the boundary face i.

4.6-2 Other Coordinate Systems

So far, we have formulated the discretization equations by using a grid in the Cartesian coordinate system. In the rest of the book, we shall continue to employ the same coordinate system for nearly all the treatment. This provides convenience of presentation and ease of understanding. However, the method being developed is not limited to Cartesian grids but can be used with a grid in any orthogonal coordinate system. To illustrate the derivation of the

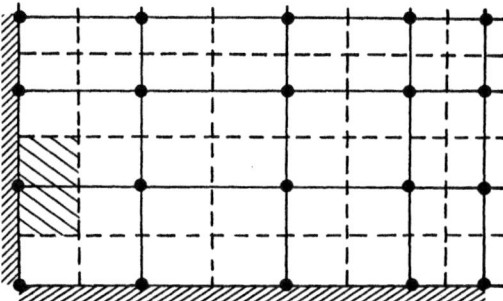

Figure 4.13 Boundary control volumes in Practice A.

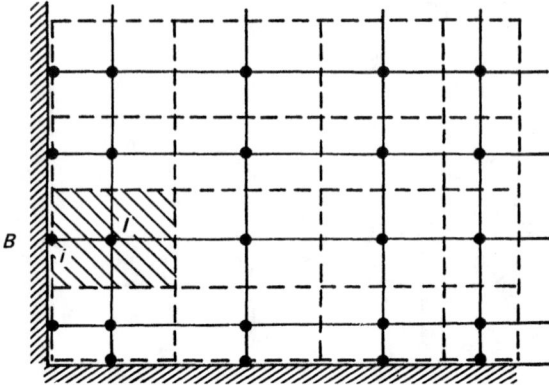

Figure 4.14 Boundary control volumes in Practice B.

discretization equation in other coordinate systems, we shall consider a two-dimensional situation in polar coordinates, namely r and θ.

The $r\theta$ counterpart of Eq. (4.42) is

$$\rho c \frac{\partial T}{\partial t} = \frac{1}{r} \frac{\partial}{\partial r} \left(rk \frac{\partial T}{\partial r} \right) + \frac{1}{r} \frac{\partial}{\partial \theta} \left(\frac{k}{r} \frac{\partial T}{\partial \theta} \right) + S . \qquad (4.57)$$

The grid and the control volume in $r\theta$ coordinates are shown in Fig. 4.15. The z-direction thickness of the control volume is assumed to be unity. To obtain the discretization equation, we multiply Eq. (4.57) by r and integrate with respect to r and θ over the control volume. (This operation gives the volume integral, since $r \, dr \, d\theta$ represents a volume element of unit thickness.)

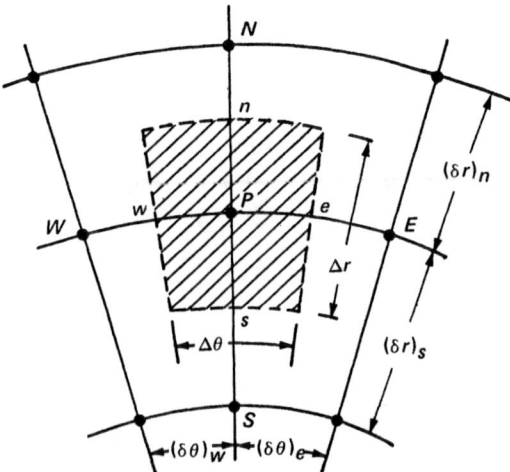

Figure 4.15 Control volume in polar coordinates.

Following the same procedure as in Section 4.4-1, we obtain the discretization equation

$$a_P T_P = a_E T_E + a_W T_W + a_N T_N + a_S T_S + b , \qquad (4.58)$$

where

$$a_E = \frac{k_e \, \Delta r}{r_e (\delta\theta)_e} , \qquad (4.59a)$$

$$a_W = \frac{k_w \, \Delta r}{r_w (\delta\theta)_w} , \qquad (4.59b)$$

$$a_N = \frac{k_n r_n \, \Delta\theta}{(\delta r)_n} , \qquad (4.59c)$$

$$a_S = \frac{k_s r_s \, \Delta\theta}{(\delta r)_s} , \qquad (4.59d)$$

$$a_P^0 = \frac{\rho c \, \Delta V}{\Delta t} , \qquad (4.59e)$$

$$b = S_C \, \Delta V + a_P^0 T_P^0 , \qquad (4.59f)$$

$$a_P = a_E + a_W + a_N + a_S + a_P^0 - S_P \, \Delta V . \qquad (4.59g)$$

Here ΔV is the volume of the control volume; it is equal to $0.5(r_n + r_s) \, \Delta\theta \, \Delta r$. (It should be noted that ΔV is not necessarily equal to $r_P \, \Delta\theta \, \Delta r$, unless P lies midway between n and s.)

The foregoing illustration shows that the additional features introduced by a new coordinate system are mainly geometric. As long as the required lengths, areas, and volumes are properly calculated, no new principles are needed. Discretization equations in any orthogonal coordinate system can now be derived along the same lines. The requirement of orthogonality, however, is essential if profiles defined by just two grid points are to be used. The fact that the control-volume face e in Fig. 4.15 is perpendicular to the line PE enables us to calculate the flux across the face from T_P and T_E alone. A more complex discretization formula would be needed for nonorthogonal grids.

In the remainder of the book, we shall use only Cartesian coordinates for all algebraic derivations. The entire treatment, however, is equally applicable to any orthogonal coordinate system when the obvious geometric changes are introduced.

4.7 CLOSURE

This chapter marks the first major step in the development of the numerical method for the general differential equation (2.13). Heat conduction presents a physical situation that embodies all the ingredients of the general equation except convection. Thus, whereas we have nearly completed the construction of the method, the remaining ingredient, namely convection, does give rise to many interesting and important considerations. The treatment of convection is not as straightforward as one would at first expect, and yet a proper treatment is crucial for handling situations with fluid flow. The next chapter is devoted to the special features that convection brings into the discretization method.

PROBLEMS

4.1 For the situation shown in Fig. 4.3, if the boundary temperature T_B were given, explain how you would obtain the heat flux q_B at the boundary, *after* the calculation of all the grid-point temperatures. (Note that an attempt to approximate dT/dx at the boundary is not consistent with the control-volume procedure; the half-control-volume equation should be used to find q_B.)

4.2 When the boundary temperature T_B in Fig. 4.3 is given, we do not use the half-control-volume equation for obtaining the temperature field. Does this mean that we do not satisfy energy conservation over the *whole* calculation domain for the given-boundary-temperature condition? (See the note for Problem 4.1.)

4.3 The boundary condition expressed by Eq. (4.19) can be thought of as the most general condition. It is then possible to obtain the two other types of boundary conditions (namely, given temperature and given heat flux) as limiting cases of this general condition. Explain how this can be achieved.

4.4 Consider the differential equation

$$\frac{d}{dx}\left(k\,\frac{dT}{dx}\right) = 0 \,.$$

Define a new variable η such that $d\eta \equiv (1/k)\,dx$. Derive the discretization equation by assuming that T is linear in η in a piecewise manner. Express η in terms of x and the grid-point conductivities by postulating that the conductivity at a grid point prevails throughout the control volume surrounding it. Verify that the resulting expression for a_E agrees with Eq. (4.11).

4.5 Derive the discretization equation from Eq. (4.1) for the situation in which $S = a + bT$, where a and b are constants. Use a piecewise-linear profile for T for calculating both dT/dx and S. Comment on the resulting discretization equation with reference to Rule 2.

4.6 Repeat the derivation in Section 4.3-1 by assuming a piecewise-linear $T \sim x$ profile also for the $\partial T/\partial t$ term. For $f = 1$ (that is, the fully implicit scheme), examine the neighbor coefficients a_E and a_W with reference to Rule 2. [Have you noticed that, with reference to Eq. (4.40), the $\partial T/\partial t$ term behaves much like S ($= S_C + S_P T_P$) and that $\partial T/\partial t$, if regarded as a part of S, would give a negative S_P as desired?]

4.7 In a combined conduction-radiation problem the source term is given by $S = a(T_0^4 - T^4)$, where a and T_0 are constants and a is positive. Write an appropriate linearization for the source term.

4.8 The source term for a dependent variable ϕ is given by $S = A - B|\phi|\phi$, where A and B are positive constants. If this term is to be linearized as $S_C + S_P\phi_P$, comment on the following practices (ϕ_P^* denotes the previous-iteration value):

(a) $S_C = A - B|\phi_P^*|\phi_P^*, \quad S_P = 0$
(b) $S_C = A, \qquad\qquad\quad S_P = -B|\phi_P^*|$
(c) $S_C = A + B|\phi_P^*|\phi_P^*, \quad S_P = -2B|\phi_P^*|$
(d) $S_C = A + 9B|\phi_P^*|\phi_P^*, \quad S_P = -10B|\phi_P^*|$

4.9 Consider a one-dimensional heat conduction situation with $S = 2$ and $k = 1$ everywhere. If four grid points at $x = 0, 1, 2, 3$ are used to span the domain of length 3, write the four discretization equations (including the half-control-volume equations) using the following boundary conditions: At $x = 0$, the heat flux *into* the domain is 5; at $x = 3$, the heat flux leaving the domain is 11.

Solve the four discretization equations by:
(a) The TDMA
(b) The Gauss-Seidel iteration
(c) Setting the temperature at the first grid point equal to 100 and applying the TDMA to the *remaining* three equations
(d) Same as (c), but solving the equations by the Gauss-Seidel method
[*Comments*: With the given boundary conditions, the values of T are not uniquely defined—the *differences* between temperatures are meaningful, but their absolute values are not. Hence, by method (a), no solution can be obtained. The solutions obtained in (b) and (c) will, in general, differ by a constant. Also, the convergence in (b) will be faster than in (d). It is, therefore, better to let the solution seek its own level than to insist on a definite value at a particular grid point.]

4.10 For the explicit scheme, Eq. (4.39) gives the stability criterion for one-dimensional problems. Derive the criteria for two- and three-dimensional situations from the requirement that the coefficient of T_P^0 must remain positive.

4.11 An infinite slab of thickness 8 units has its faces maintained at a temperature of 100. The temperature field is governed by Eq. (4.1) with $k = 5$ and $S = 50$ everywhere. Using only a few grid points, obtain a numerical solution by the method developed in this chapter. Compare the values of T from the solution with those from the exact solution. (If the grid is designed according to Practice A, the agreement with the exact solution will be perfect. Why?)

4.12 Formulate the following problem in terms of appropriate dimensionless variables: The governing equation is

$$k\,\frac{d^2 T}{dx^2} + S = 0\,,$$

where k and S are constant. The boundary conditions are

$$x = 0 \qquad -k\,\frac{dT}{dx} = h_0(T_f - T_0)\,,$$

$$x = L \qquad -k\,\frac{dT}{dx} = h_L(T_L - T_f)\,,$$

where h_0 and h_L are the heat transfer coefficients, and T_0 and T_L are the corresponding boundary temperatures. Solve the problem numerically for the case $h_0 L/k = 1$ and $h_L L/k = 2$, and compare the results with the exact solution.

4.13 A number of simple fully developed flows are governed by conductionlike equations. For example, the fully developed flow between parallel plates obeys the equation

$$\frac{d}{dy}\left(\mu\,\frac{du}{dy}\right) - \frac{dp}{dx} = 0\,,$$

where u is the velocity, μ is the viscosity, and dp/dx is the constant pressure gradient. Noting that this equation is essentially identical to Eq. (4.1), we can use the discretization method developed in this chapter for calculating fully developed flows.

(a) Compute the velocity distribution in the fully developed flow between stationary parallel plates.

(b) Let one of the plates be stationary, while the other is moving with velocity U. Calculate the fully developed flow between the plates for various values of the parameter $L^2 (dp/dx)/(\mu U)$, where L is the distance between the two plates.

(c) Calculate the velocity field for the fully developed flow in a circular pipe.

4.14 The thermally fully developed region in a duct is characterized by a temperature field that, when expressed in appropriate dimensionless form, remains unchanged with the streamwise distance. Calculate the fully developed temperature field and the Nusselt number in a fully developed flow between two parallel plates, assuming that the velocity profile is parabolic, one plate is adiabatic, and there is a uniform heat flux across the other plate. [A large variety of fully developed flow and heat transfer problems can now be solved by the method developed in this chapter. You may wish to verify some of the results presented in Sparrow and Patankar (1977).]

4.15 Consider unsteady heat conduction in an infinite slab. One face of the slab is insulated, while a constant heat flux enters the slab through the other face. After the initial transient, the temperature profile will acquire a fixed shape, and all the temperatures will rise with time at the same rate. Further, this rate will be related to the amount of heat flux through the face. Formulate and solve the problem by the techniques of *steady-state* heat conduction. [Such "fully developed" regime in unsteady heat conduction is discussed more fully in Patankar (1979b).]

4.16 Consider the one-dimensional heat conduction problem in a rod that is bent into a circular shape to form an endless loop. It thus has no exposed ends and no meaningful boundary conditions. Indeed, all grid points will be the internal grid points. The discretization equations will still have the form (4.22), but the conditions given by Eq. (4.23) will not apply. Instead, T_{N+1} will be interpreted as T_1, and T_0 as T_N. Derive a solution algorithm (which we shall call the circular TDMA) for such a set of equations. [This algorithm will be useful in applying the line-by-line method in $r\theta$ coordinates, because the grid points forming a θ-direction line may be arranged in an endless loop. Another application of the circular TDMA, and the details of its derivation, can be found in Patankar, Liu, and Sparrow (1977).]

4.17 Consider two dependent variables f and g, which are governed by coupled equations of the form

$$a_i f_i = b_i f_{i+1} + c_i f_{i-1} + d_i + e_i g_i \ ,$$

and

$$A_i g_i = B_i g_{i+1} + C_i g_{i-1} + D_i + E_i f_i \ ,$$

for $i = 1, 2, 3, \ldots, N$. Also, $c_1 = 0$, $b_N = 0$, $C_1 = 0$, and $B_N = 0$. Using the basic ideas of the TDMA, derive an algorithm for solving these equations.

4.18 Compare Eqs. (4.56) and (4.55b) to show that the inertia i that is implicit in the use of a relaxation factor α is given by $i = (1 - \alpha)a_P/\alpha$.

4.19 A slab of thickness L has a linear temperature distribution within it from $T = T_0$ at $x = 0$ to $T = T_1$ at $x = L$. At time $t = 0$, the face at $x = L$ is made adiabatic, while the face at $x = 0$ is still held at $T = T_0$. Calculate the distribution of $(T - T_0)/(T_1 - T_0)$ as a function of x/L and $\alpha t/L^2$, where α is the thermal diffusivity. Continue the computations until the value of $(T - T_0)/(T_1 - T_0)$ at $x = L$ falls below 0.5.

4.20 Consider the steady one-dimensional conduction in a constant-area fin governed by

$$\frac{d}{dx} \left(k \frac{dT}{dx} \right) + \frac{hP}{A} (T_f - T) = 0 \,,$$

where h is the heat transfer coefficient between the fin surface and the surrounding fluid at temperature T_f, A is the cross-sectional area of the fin, and P is the perimeter of the cross section. The boundary conditions are: At $x = 0$, $T = T_0$ (the base temperature), and at $x = L$, $k \, dT/dx = 0$ (insulated tip). Find the numerical solution for the dimensionless temperature $(T - T_f)/(T_0 - T_f)$ as a function of x/L for $hPL^2/kA = 2$, and compare it with the exact solution. For a uniform grid, find the number of grid points needed to predict the heat flux at the base within 1% of the exact value. (Note that the proper linearization of the source term in the given equation is quite obvious. However, if you attempt to solve the problem iteratively by expressing the entire source term as S_C and setting $S_P = 0$, you will observe that the iterations successively produce unrealistic results and make the convergence difficult to attain.)

FIVE

CONVECTION AND DIFFUSION

5.1 THE TASK

So far, in the guise of heat conduction, we have seen how to formulate the discretization equation from the general differential equation containing the unsteady term, the diffusion term, and the source term. (The description in the last chapter in terms of temperature T and conductivity k can easily be recast in terms of the general variable ϕ and its diffusion coefficient Γ.) The only omission has been the convection term, which we shall now include. We have also dealt with the methods of solving the algebraic equations; as long as the addition of the convection term does not alter the *form* of the discretization equation, the same methods continue to apply.

The convection is created by fluid flow. Our task in this chapter is to obtain a solution for ϕ in the presence of a *given flow field* (i.e., the velocity components and the density). How we know the flow field is a question we do not ask at this stage. It could have come from experiment, be given as an analytical solution, be obtained by the method described later in Chapter 6, or simply be guessed. The origin of the flow-field information is immaterial here. Having somehow acquired the flow field, we wish to calculate the temperature, concentration, enthalpy, or any such quantity that is represented by the general variable ϕ.

Although convection is the only new term introduced in this chapter, its formulation is not very straightforward. The convection term has an in-separable connection with the diffusion term, and therefore, the two terms need to be handled as one unit. This is why the words "convection and diffusion" form the title of this chapter; other terms can also be present, but only in the background.

It should be remembered that the word diffusion is used here in a generalized sense. It is not restricted only to the diffusion of a chemical species caused by concentration gradients. The diffusion flux due to the gradient of the general variable ϕ is $-\Gamma\ \partial\phi/\partial x_j$, which, for specific meanings of ϕ, would represent chemical-species diffusion flux, heat flux, viscous stress, etc. The general differential equation (2.15) contains the term $(\partial/\partial x_j)\ (\Gamma\ \partial\phi/\partial x_j)$, which is designated as the diffusion term. Actually, this expression denotes the sum of three terms for the three coordinate directions; yet it is convenient to refer to them collectively as the diffusion term. The same is true of the convection term, which is $(\partial/\partial x_j)\ (\rho u_j \phi)$.

One feature of the convection-diffusion situation may be noted at this point. Since the given flow field must satisfy the continuity equation

$$\frac{\partial \rho}{\partial t} + \frac{\partial}{\partial x_j}\ (\rho u_j) = 0 \ , \tag{5.1}$$

the general differential equation

$$\frac{\partial}{\partial t}\ (\rho\phi) + \frac{\partial}{\partial x_j}\ (\rho u_j \phi) = \frac{\partial}{\partial x_j}\left(\Gamma\ \frac{\partial\phi}{\partial x_j} \right) + S \tag{5.2}$$

can also be written as

$$\rho\ \frac{\partial\phi}{\partial t} + \rho u_j\ \frac{\partial\phi}{\partial x_j} = \frac{\partial}{\partial x_j}\left(\Gamma\ \frac{\partial\phi}{\partial x_j} \right) + S \ . \tag{5.3}$$

From this form of the equation, it follows that, for given distributions of ρ, u_j, Γ, and S, any solution ϕ and its variant (ϕ plus a constant) would both satisfy Eq. (5.3). Under these circumstances, the basic rule about the sum of the coefficients (Rule 4) continues to apply.

5.2 STEADY ONE–DIMENSIONAL CONVECTION AND DIFFUSION

As in the last chapter, much can be learned from consideration of the simplest possible case. Here we shall consider a steady one-dimensional situation in which only the convection and diffusion terms are present. The governing differential equation is

$$\frac{d}{dx}\ (\rho u\phi) = \frac{d}{dx}\left(\Gamma\ \frac{d\phi}{dx} \right) \ , \tag{5.4}$$

where u represents the velocity in the x direction. Also, the continuity equation becomes

$$\frac{d}{dx}(\rho u) = 0 \qquad \text{or} \qquad \rho u = \text{constant} . \qquad (5.5)$$

For deriving the discretization equation, we shall use the three-grid-point cluster shown in Fig. 5.1. Although the actual location of the control-volume faces e and w would not influence our final formulation, it is convenient to assume that e is located *midway* between P and E, and w midway between W and P.

5.2-1 A Preliminary Derivation

Integration of Eq. (5.4) over the control volume shown in Fig. 5.1 gives

$$(\rho u \phi)_e - (\rho u \phi)_w = \left(\Gamma \frac{d\phi}{dx}\right)_e - \left(\Gamma \frac{d\phi}{dx}\right)_w . \qquad (5.6)$$

We saw in the last chapter how to represent the term $\Gamma\, d\phi/dx$ from a piecewise-linear profile for ϕ. For the convection term, the same choice of profile would at first seem natural. The result is

$$\phi_e = \tfrac{1}{2}(\phi_E + \phi_P) \qquad \text{and} \qquad \phi_w = \tfrac{1}{2}(\phi_P + \phi_W) . \qquad (5.7)$$

The factor $\tfrac{1}{2}$ arises from the assumption of the interfaces being midway; some other interpolation factors would have appeared for differently located interfaces. Now, Eq. (5.6) can be written as

$$\tfrac{1}{2}(\rho u)_e(\phi_E + \phi_P) - \tfrac{1}{2}(\rho u)_w(\phi_P + \phi_W) = \frac{\Gamma_e(\phi_E - \phi_P)}{(\delta x)_e} - \frac{\Gamma_w(\phi_P - \phi_W)}{(\delta x)_w} ,$$
$$(5.8)$$

where the values of Γ_e and Γ_w are to be obtained by the prescription presented in Section 4.2-3. (This applies throughout the book, although such references to previous sections may not be repeated.)

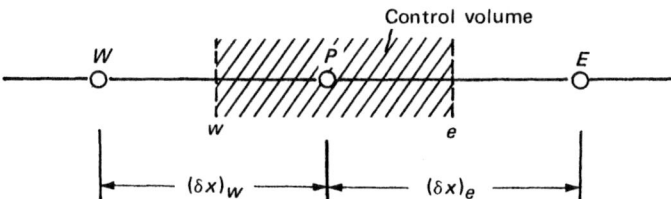

Figure 5.1 Typical grid-point cluster for the one-dimensional problem.

To arrange the equation more compactly, we define two new symbols F and D, as follows:

$$F \equiv \rho u, \qquad D \equiv \frac{\Gamma}{\delta x} . \tag{5.9}$$

Both have the same dimensions; F indicates the strength of the convection (or flow), while D is the diffusion conductance. It should be noted that, whereas D always remains positive, F can take either positive or negative values depending on the direction of the fluid flow. With the new symbols, the discretization equation becomes

$$a_P \phi_P = a_E \phi_E + a_W \phi_W , \tag{5.10}$$

where

$$a_E = D_e - \frac{F_e}{2} , \tag{5.11a}$$

$$a_W = D_w + \frac{F_w}{2} , \tag{5.11b}$$

$$a_P = D_e + \frac{F_e}{2} + D_w - \frac{F_w}{2}$$

$$= a_E + a_W + (F_e - F_w) . \tag{5.11c}$$

Discussion. (1) Since by continuity $F_e = F_w$, we do get the property $a_P = a_E + a_W$. Further, it is interesting to note from Eq. (5.11c) that the discretization equation has this property only if the flow field satisfies continuity, just as Eq. (5.3) can be derived from Eq. (5.2) only if the continuity equation is satisfied. (2) The discretization equation (5.10) represents the implications of the piecewise-linear profile for ϕ. This form is also known as the *central-difference* scheme and is the natural outcome of a Taylor-series formulation. (3) It is instructive to consider a simple example in which

$$D_e = D_w = 1 \qquad \text{and} \qquad F_e = F_w = 4 .$$

Further, if the values of ϕ_E and ϕ_W are given, we can obtain ϕ_P from Eq. (5.10). Consider two sets of values:

(*a*) If $\phi_E = 200$ and $\phi_W = 100$, the result is $\phi_P = 50$!
(*b*) If $\phi_E = 100$ and $\phi_W = 200$, the result is $\phi_P = 250$!

Since ϕ_P, in reality, cannot fall outside the range of 100–200 established by its neighbors, these results are clearly unrealistic. (4) Indeed, we could have anticipated these unrealistic results, because Eqs. (5.11) indicate that the coefficients could, at times, become negative. When $|F|$ exceeds $2D$, then, depending on whether F is positive or negative, there is a possiblity of a_E or a_W becoming negative. This will be a violation of one of the basic rules, with a possible disastrous outcome. (5) Also, the negative coefficients would imply that a_P, which equals $'\Sigma a_{nb}$, is less than $\Sigma |a_{nb}|$, which fails to satisfy the Scarborough criterion. Then, a point-by-point solution of the discretization equations may diverge. This is why all the early attempts to solve convective problems by the central-difference scheme were limited to low Reynolds numbers (i.e., to low values of F/D). (6) For the case of zero diffusion (that is, $\Gamma = 0$), the scheme leads to $a_P = 0$. Then, Eq. (5.10) becomes unsuitable for solution by a point-by-point method, and by most other iterative methods.

Since the foregoing preliminary formulation has resulted in an unacceptable discretization equation, we must seek better formulations. Some such possibilities are described in the following subsections.

5.2-2 The Upwind Scheme

A well-known remedy for the difficulties encountered is the *upwind* scheme, which is also known as the upwind-difference scheme, the upstream-difference scheme, the donor-cell method, etc. It was first put forward by Courant, Isaacson, and Rees (1952) and subsequently reinvented by Gentry, Martin, and Daly (1966), Barakat and Clark (1966), and Runchal and Wolfshtein (1969).

The upwind scheme recognizes that the weak point in the preliminary formulation is the assumption that the convected property ϕ_e at the interface is the average of ϕ_E and ϕ_P, and it proposes a better prescription. The formulation of the diffusion term is left unchanged, but the convection term is calculated from the following assumption:

The value of ϕ at an interface is equal to the value of ϕ at the grid point on the *upwind* side of the face.

Thus,

$$\phi_e = \phi_P \qquad \text{if} \qquad F_e > 0 \,, \qquad (5.12a)$$

and $$\phi_e = \phi_E \qquad \text{if} \qquad F_e < 0 \,. \qquad (5.12b)$$

The value of ϕ_w can be defined similarly.

The conditional statements (5.12) can be more compactly written if we define a new operator.* We shall define $[\![A, B]\!]$ to denote the greater of A and B. Then, the upwind scheme implies

$$F_e \phi_e = \phi_P [\![F_e, 0]\!] - \phi_E [\![-F_e, 0]\!] . \qquad (5.13)$$

When Eq. (5.7) is replaced by this concept, the discretization equation becomes

$$a_P \phi_P = a_E \phi_E + a_W \phi_W , \qquad (5.14)$$

where

$$a_E = D_e + [\![-F_e, 0]\!] , \qquad (5.15a)$$

$$a_W = D_w + [\![F_w, 0]\!] , \qquad (5.15b)$$

$$a_P = D_e + [\![F_e, 0]\!] + D_w + [\![-F_w, 0]\!]$$

$$= a_E + a_W + (F_e - F_w) . \qquad (5.15c)$$

Discussion. (1) It is evident from Eqs. (5.15) that no negative coefficients would arise. Thus, the solutions will always be physically realistic, and the Scarborough criterion will be satisfied. (2) What is, however, the rationale for the main idea underlying the upwind scheme? More insight will be obtained in the next subsection, but, in the meantime, a lucid physical picture of the upwind scheme would offer some satisfaction. The scheme is sometimes said to be based on the "tank-and-tube" model (Gosman, Pun, Runchal, Spalding, and Wolfshtein, 1969). As shown in Fig. 5.2, the control volumes can be thought to be stirred tanks that are connected in series by short tubes. The flow through the tubes represents convection, while the conduction through the tank walls represents diffusion. Since the tanks are stirred, each contains a uniform temperature fluid. Then, it is appropriate to suppose that the fluid flowing in each connecting tube has the temperature that prevails in the tank

---

*This new operator $[\![A, B]\!]$ is equivalent to AMAX1(A, B) in the computer language FORTRAN.

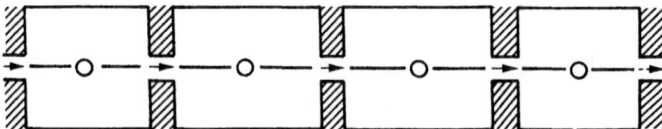

Figure 5.2 Tank-and-tube model.

on the upstream side. Normally, the fluid in the tube would not know anything about the tank toward which it is heading, but would carry the full legacy of the tank from which it has come. This is the essence of the upwind scheme.

5.2-3 The Exact Solution

Fortunately, the governing equation (5.4) can be solved exactly if Γ is taken to be constant [ρu is already constant, as given by Eq. (5.5)]. If a domain $0 \leqslant x \leqslant L$ is used, with the boundary conditions

$$\text{At} \quad x = 0 \quad \phi = \phi_0 , \qquad (5.16a)$$

$$\text{At} \quad x = L \quad \phi = \phi_L , \qquad (5.16b)$$

the solution of Eq. (5.4) is

$$\frac{\phi - \phi_0}{\phi_L - \phi_0} = \frac{\exp (Px/L) - 1}{\exp (P) - 1} , \qquad (5.17)$$

where P is a Peclet number defined by

$$P \equiv \frac{\rho u L}{\Gamma} . \qquad (5.18)$$

It can be seen that P is the ratio of the strengths of convection and diffusion.

The nature of the exact solution (5.17) can be understood from Fig. 5.3 where the $\phi \sim x$ variation has been plotted for different values of the Peclet number. In the limit of zero Peclet number, we get the pure-diffusion (or conduction) problem, and the $\phi \sim x$ variation is linear. When the flow is in the positive x direction (i.e., for positive values of P), the values of ϕ in the domain seem to be more influenced by the upstream value ϕ_0. For a large positive value of P, the value of ϕ remains very close to the upstream value ϕ_0 over much of the domain. The picture is reversed for negative values of P. When the fluid flows in the negative x direction, ϕ_L becomes the *upstream* value, which dominates the values of ϕ in the domain. For a large negative P, the value of ϕ over most of the region is very nearly equal to ϕ_L.

Implications. For constructing the discretization equation, we can now obtain guidance from Fig. 5.3 regarding the appropriate $\phi \sim x$ profile between grid points. (1) It is easy to see why our preliminary derivation failed to give a satisfactory formulation. The $\phi \sim x$ profile is far from being linear except for small values of $|P|$. (2) When $|P|$ is large, the value of ϕ at $x = L/2$ (the interface) is nearly equal to the value of ϕ at the upwind boundary. This is precisely the assumption made in the upwind scheme; but there it is used for

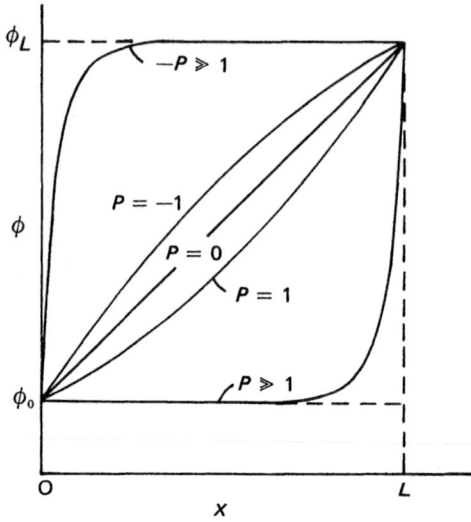

Figure 5.3 Exact solution for the one-dimensional convection-diffusion problem.

all values of $|P|$, not just for large values. (3) When $|P|$ is large, $d\phi/dx$ is nearly zero at $x = L/2$. Thus, the diffusion is almost absent. The upwind scheme always calculates the diffusion term from a linear $\phi \sim x$ profile and thus overestimates diffusion at large values of $|P|$.

If the discretization equation were to be obtained directly from the exact solution shown in Fig. 5.3, the resulting scheme would not have any of these defects. Let us proceed to derive such a scheme, which we shall name the *exponential* scheme. It is based on the formulation first presented by Spalding (1972) and is one of the schemes proposed and employed by Raithby and Torrance (1974).

5.2-4 The Exponential Scheme

It is useful to consider a total flux J that is made up of the convection flux $\rho u\phi$ and the diffusion flux $-\Gamma \, d\phi/dx$. Thus,

$$J = \rho u\phi - \Gamma \frac{d\phi}{dx} . \tag{5.19}$$

With this definition, Eq. (5.4) becomes

$$\frac{dJ}{dx} = 0 , \tag{5.20}$$

which, when integrated over the control volume shown in Fig. 5.1, gives

$$J_e - J_w = 0 . \tag{5.21}$$

Now the exact solution (5.17) can be used as a profile between points P and E, with ϕ_P and ϕ_E replacing ϕ_0 and ϕ_L, and the distance $(\delta x)_e$ replacing L. The substitution of this profile into Eq. (5.19) would give the expression for J_e:

$$J_e = F_e \left(\phi_P + \frac{\phi_P - \phi_E}{\exp{(P_e)} - 1} \right) , \tag{5.22}$$

where

$$P_e = \frac{(\rho u)_e (\delta x)_e}{\Gamma_e} = \frac{F_e}{D_e} , \tag{5.23}$$

and F_e and D_e are as defined* by Eq. (5.9). It should be noted that J_e does not depend on the location of the interface between points P and E. Of course, an exact solution that obeys Eq. (5.20) must exhibit this behavior.

Finally, substitution of Eq. (5.22) and a similar expression for J_w into Eq. (5.21) leads to

$$F_e \left(\phi_P + \frac{\phi_P - \phi_E}{\exp{(P_e)} - 1} \right) - F_w \left(\phi_W + \frac{\phi_W - \phi_P}{\exp{(P_w)} - 1} \right) = 0 , \tag{5.24}$$

which can be cast into our standard form

$$a_P \phi_P = a_E \phi_E + a_W \phi_W , \tag{5.25}$$

where

$$a_E = \frac{F_e}{\exp{(F_e/D_e)} - 1} , \tag{5.26a}$$

$$a_W = \frac{F_w \exp{(F_w/D_w)}}{\exp{(F_w/D_w)} - 1} , \tag{5.26b}$$

$$a_P = a_E + a_W + (F_e - F_w) . \tag{5.26c}$$

*Here Γ_e is to be obtained in the same manner as k_e was derived in Eq. (4.9). This may seem like a neat way in which the exact solution for constant Γ is boldly modified to accept a nonuniform Γ. Although there would be no objection to such a practice, the prescription for k_e given by Eq. (4.9) (which was derived for the conduction situation) happens to be the exact formula for Γ_e even in the convection-diffusion case (see Problem 5.5).

These coefficient expressions define the exponential scheme. When used for the steady one-dimensional problem, this scheme is guaranteed to produce the exact solution for any value of the Peclet number and for any number of grid points. Despite its highly desirable behavior, it is not widely used because (1) exponentials are expensive to compute, and (2) since the scheme is not exact for two- or three-dimensional situations, nonzero sources, etc., the extra expense of computing the exponentials does not seem to be justified.

What we really need is an easy-to-compute scheme that has the qualitative behavior of the exponential scheme. Two such schemes will now be presented; the second of these is recommended for use.

5.2-5 The Hybrid Scheme

The hybrid scheme was developed by Spalding (1972); it also appears in the book by Patankar and Spalding (1970) under the name "high-lateral-flux modification."

To appreciate the connection between the exponential scheme and the hybrid scheme, we shall plot the coefficient a_E, or rather its dimensionless form a_E/D_e, as a function of the Peclet number P_e. From Eq. (5.26) we deduce that

$$\frac{a_E}{D_e} = \frac{P_e}{\exp{(P_e)} - 1} \, . \tag{5.27}$$

The variation of a_E/D_e with P_e is shown in Fig. 5.4. For positive values of P_e, the grid point E is the *downstream* neighbor, and its influence is seen to decrease as P_e increases. When P_e is negative, E is the *upstream* neighbor and has a large influence. Certain specific properties of the exact variation of a_E/D_e (shown by the solid line in Fig. 5.4) can be seen to be:

1. For $P_e \to \infty$,

$$\frac{a_E}{D_e} \to 0 \, ; \tag{5.28a}$$

2. For $P_e \to -\infty$,

$$\frac{a_E}{D_e} \to -P_e \, ; \tag{5.28b}$$

3. At $P_e = 0$, the tangent is

$$\frac{a_E}{D_e} = 1 - \frac{P_e}{2} \, . \tag{5.28c}$$

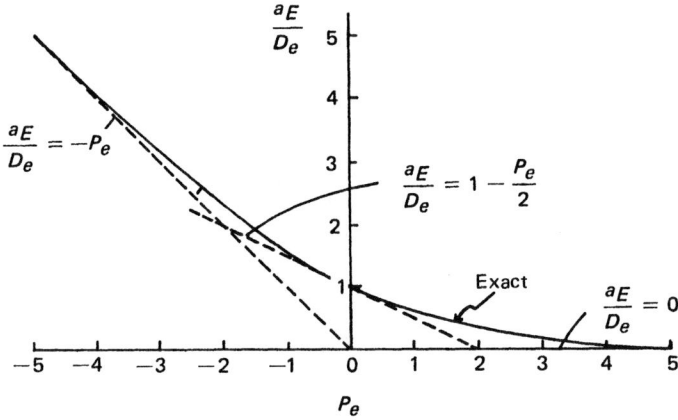

Figure 5.4 Variation of the coefficient a_E with Peclet number.

The three straight lines representing these limiting cases are also shown in Fig. 5.4. They can be seen to form an envelope of, and represent a reasonable approximation to, the exact curve. The hybrid scheme is indeed made up of these three straight lines, so that

For $P_e < -2$,

$$\frac{a_E}{D_e} = -P_e ,\qquad\qquad (5.29a)$$

For $-2 \leqslant P_e \leqslant 2$,

$$\frac{a_E}{D_e} = 1 - \frac{P_e}{2},\qquad\qquad (5.29b)$$

For $P_e > 2$,

$$\frac{a_E}{D_e} = 0.\qquad\qquad (5.29c)$$

These expressions can be combined into a compact form by the use of the special symbol $[\![\ \ ]\!]$, which stands for the largest of the quantities contained within it. Thus,

$$a_E = D_e \left[\!\left[-P_e,\ 1 - \frac{P_e}{2},\ 0 \right]\!\right] ,\qquad\qquad (5.30a)$$

or

$$a_E = \left[\!\left[-F_e,\ D_e - \frac{F_e}{2},\ 0 \right]\!\right] .\qquad\qquad (5.30b)$$

The significance of the hybrid scheme can be understood by observing that (1) it is identical with the central-difference scheme for the Peclet-number range $-2 \leqslant P_e \leqslant 2$, and (2) outside this range it reduces to the upwind scheme in which the diffusion has been set equal to zero. Thus, the shortcomings of the upwind scheme listed at the end of Section 5.2-3 are not shared by the hybrid scheme. The name hybrid is indicative of a combination of the central-difference and upwind schemes, but it is best to consider the hybrid scheme as the three-line approximation to the exact curve, as shown in Fig. 5.4.

The convection-diffusion discretization equation for the hybrid scheme can now be written as

$$a_P \phi_P = a_E \phi_E + a_W \phi_W , \qquad (5.31)$$

where

$$a_E = \left[\!\left[-F_e, D_e - \frac{F_e}{2}, 0 \right]\!\right] , \qquad (5.32a)$$

$$a_W = \left[\!\left[F_w, D_w + \frac{F_w}{2}, 0 \right]\!\right] , \qquad (5.32b)$$

$$a_P = a_E + a_W + (F_e - F_w) . \qquad (5.32c)$$

It should be remembered that this formulation is valid for any arbitrary location of the interfaces between the grid points and is *not* limited to midway interfaces.

5.2-6 The Power-Law Scheme

It can be seen from Fig. 5.4 that the departure of the hybrid scheme from the exact curve is rather large at $P_e = \pm 2$; also, it seems rather premature to set the diffusion effects equal to zero as soon as $|P_e|$ exceeds 2. A better approximation to the exact curve is given by the power-law scheme, which is described in Patankar (1979a). Although somewhat more complicated than the hybrid scheme, the power-law expressions are not particularly expensive to compute, and they provide an extremely good representation of the exponential behavior.

The power-law expressions for a_E can be written as

For $P_e < -10$,

$$\frac{a_E}{D_e} = -P_e , \qquad (5.33a)$$

For $-10 \leqslant P_e < 0$,

$$\frac{a_E}{D_e} = (1 + 0.1P_e)^5 - P_e, \qquad (5.33b)$$

For $0 \leqslant P_e \leqslant 10$,

$$\frac{a_E}{D_e} = (1 - 0.1P_e)^5, \qquad (5.33c)$$

For $P_e > 10$,

$$\frac{a_E}{D_e} = 0. \qquad (5.33d)$$

Comparing these expressions with Eqs. (5.29), we observe that, for $|P_e| > 10$, the power-law scheme becomes identical with the hybrid scheme. A compact form for Eqs. (5.33) can be written as

$$a_E = D_e \left[\!\!\left[0, \left(1 - \frac{0.1|F_e|}{D_e}\right)^5 \right]\!\!\right] + [\!\![0, -F_e]\!\!] . \qquad (5.34)$$

The closeness of the power-law scheme to the exact exponential scheme can be judged from Table 5.1; the difference between the two schemes is too

Table 5.1 Comparison of coefficient values
given by power-law and exponential schemes

P_e	Values of a_E/D_e	
	Power-law scheme	Exponential scheme
−20	20.00	20.00
−10	10.00	10.00
−5	5.031	5.034
−4	4.078	4.075
−3	3.168	3.157
−2	2.328	2.313
−1	1.590	1.582
−0.5	1.274	1.271
0	1	1
0.5	0.7738	0.7707
1	0.5905	0.5820
2	0.3277	0.3130
3	0.1681	0.1572
4	0.07776	0.07463
5	0.03125	0.03392
10	0	0.00045
20	0	4.1×10^{-8}

small for a useful graphical comparison. As mentioned before, the power-law scheme is the recommended convection-diffusion formulation in this book, although the hybrid scheme should serve just as well in many situations.

5.2-7 A Generalized Formulation

To obtain further insight into the convection-diffusion formulation and to construct a general framework into which the various schemes considered so far can be fitted, we shall now explore some general properties of the coefficients involved. Let us consider the grid points i and $i + 1$ separated by a distance δ, as shown in Fig. 5.5. We are interested in representing the *total* flux J crossing an interface between these grid points. By use of Eq. (5.19), we write

$$J^* \equiv \frac{J\delta}{\Gamma} = P\phi - \frac{d\phi}{d(x/\delta)} \,, \tag{5.35}$$

where P is the Peclet number, $\rho u \delta / \Gamma$. The value of ϕ at the interface will be some weighted average of ϕ_i and ϕ_{i+1}, while the gradient $d\phi/d(x/\delta)$ will be some multiple of $\phi_{i+1} - \phi_i$. Thus, we propose

$$J^* = P[\alpha\phi_i + (1 - \alpha)\phi_{i+1}] - \beta(\phi_{i+1} - \phi_i) \,, \tag{5.36}$$

where α and β are dimensionless multipliers that depend on P. In this manner, J^* can be expressed as

$$J^* = B\phi_i - A\phi_{i+1} \,, \tag{5.37}$$

where A and B are dimensionless coefficients that are functions of the Peclet number P. (The coefficient A is associated with the grid point $i + 1$, which is *A*head of the interface, while B is connected with the grid point i, which is *B*ehind the interface, as seen from the chosen coordinate direction.)

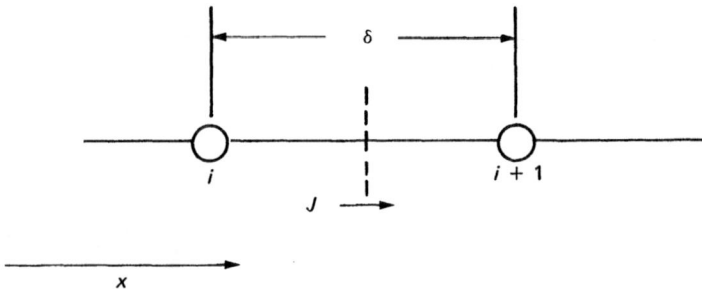

Figure 5.5 Total flux J between two grid points.

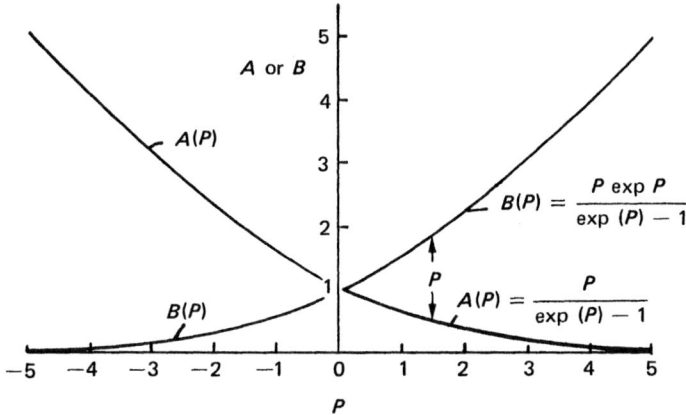

Figure 5.6 Variation of A and B with Peclet number.

Properties of A *and* B. Two properties of the coefficients A and B are particularly useful in studying their dependence on the Peclet number. First, we note that if ϕ_i and ϕ_{i+1} are equal, the diffusion flux must be zero, and J would then simply be the convection flux $\rho u \phi_i$. Thus, under these conditions, we have

$$J^* = P\phi_i = P\phi_{i+1} \, . \tag{5.38}$$

Combination of Eqs. (5.37) and (5.38) leads* to

$$B = A + P \, . \tag{5.39}$$

The second property of A and B is a kind of symmetry between them. If we reverse the coordinate axis, then P will appear as $-P$, and A and B will interchange their roles. Thus, the functions $A(P)$ and $B(P)$ must be related by

$$A(-P) = B(P) \tag{5.40a}$$

or

$$B(-P) = A(P) \, . \tag{5.40b}$$

Implications of the properties. The exact variation of A and B with the Peclet number P, which can be deduced from Eq. (5.22), is shown in Fig. 5.6, where the aforementioned properties can be observed. The vertical distance between the A and B curves can be seen to be equal to P; also, the two curves exhibit symmetry about the $P = 0$ location. The main implication of the two

---

*Alternatively, from Eqs. (5.36) and (5.37), we obtain $B = P\alpha + \beta$ and $A = P\alpha + \beta - P$; these expressions lead to the relationship stated in Eq. (5.39).

properties is that the complete $A(P)$ and $B(P)$ functions can be specified once the function $A(P)$ for only positive values of P is known (i.e., the curve shown as a thick line in Fig. 5.6). This follows since, for $P < 0$,

$$A(P) = B(P) - P \qquad \text{from (5.39)}$$

$$= A(-P) - P \qquad \text{from (5.40a)}$$

$$= A(|P|) - P . \qquad (5.41)$$

Thus, for all values of P, positive and negative, we can write

$$A(P) = A(|P|) + [\![-P, 0]\!] , \qquad (5.42)$$

and then, by use of Eq. (5.39), we get

$$B(P) = A(|P|) + [\![P, 0]\!] . \qquad (5.43)$$

Also, we shall record here, for future use, the following two relations obtained by combining Eqs. (5.37) and (5.39):

$$J^* - P\phi_i = A(\phi_i - \phi_{i+1}) , \qquad (5.44)$$

$$J^* - P\phi_{i+1} = B(\phi_i - \phi_{i+1}) . \qquad (5.45)$$

If we now apply the flux relationship (5.37) to the interfaces e and w and use Eqs. (5.42) and (5.43), we obtain the following general convection-diffusion formulation:

$$a_P \phi_P = a_E \phi_E + a_W \phi_W , \qquad (5.46)$$

where

$$a_E = D_e A(|P_e|) + [\![-F_e, 0]\!] , \qquad (5.47a)$$

$$a_W = D_w A(|P_w|) + [\![F_w, 0]\!] , \qquad (5.47b)$$

$$a_P = a_E + a_W + (F_e - F_w) . \qquad (5.47c)$$

The various schemes derived so far can now be thought of as merely different choices of the function $A(|P|)$. Expressions for $A(|P|)$ for the schemes considered so far are listed in Table 5.2 and shown graphically in Fig. 5.7. The degree of satisfactoriness of each function can be judged by comparison with the exact function.

Table 5.2 The function $A(|P|)$ for different schemes

Scheme	Formula for $A(|P|)$
Central difference	$1 - 0.5|P|$
Upwind	1
Hybrid	$[\![0, 1 - 0.5|P|]\!]$
Power law	$[\![0, (1 - 0.1|P|)^5]\!]$
Exponential (exact)	$|P|/[\exp(|P|) - 1]$

5.2-8 Consequences of the Various Schemes

Before leaving the one-dimensional problem, we shall examine the values of ϕ_P predicted by the various schemes for given values of ϕ_E and ϕ_W. Let us set, without loss of generality, the values $\phi_E = 1$ and $\phi_W = 0$. Further, let the distances $(\delta x)_e$ and $(\delta x)_w$ be equal; then ϕ_P will be a function of P ($\equiv \rho u \delta x / \Gamma$). The values of ϕ_P given by the different schemes for various values of P are shown in Fig. 5.8. (The results of the power-law scheme and the exact solution are too close to be plotted as separate curves.) All schemes except the central-difference scheme give what may be termed a physically realistic solution; the central-difference scheme, on the other hand, produces some values that lie outside the 0-1 range established by the boundary values.

Since it is the *grid* Peclet number that decides the behavior of the numerical schemes, it is, in principle, possible to refine the grid (i.e., to use smaller δx) until P is small enough (< 2) for the central-difference scheme to yield reasonable solutions. In most practical problems, however, this strategy

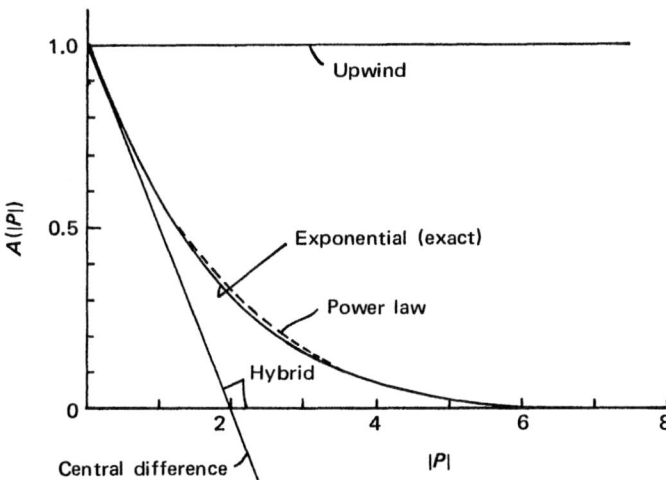

Figure 5.7 The function $A(|P|)$ for various schemes.

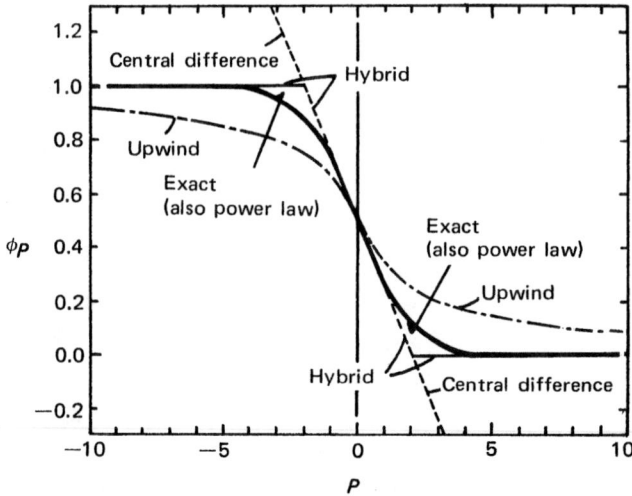

Figure 5.8 Prediction of ϕ_P by the various schemes for a range of Peclet numbers.

requires excessively fine grids, which are usually not feasible on economic grounds; in any case, we could not accept such a constraint while seeking procedures that would give physically realistic solutions even for coarse grids.

5.3 DISCRETIZATION EQUATION FOR TWO DIMENSIONS

Now we have all the ingredients needed for writing the discretization equation corresponding to the general differential equation (5.2). At first, we shall derive only the two-dimensional form, but the same procedure would apply to three dimensions.

Let us consider the control volume shown in Fig. 5.9. If we employ our one-dimensional practice of obtaining the total flux J_e, and assume that it prevails over the control-volume face of area $\Delta y \times 1$, we shall be in a position to write the complete discretization equation at once. This is given in Section 5.3-2, to which the reader with no need for the finer details of the derivation may safely jump.

5.3-1 Details of the Derivation

One subtle detail of the derivation will now be given some attention. Even in the one-dimensional situation we have seen that a_P turned out to be $a_E + a_W$ only when the continuity equation was satisfied. Thus, our basic rule about the sum of the neighbor coefficients (Rule 4) can be satisfied only when we involve the continuity equation in the derivation. This practice is illustrated in the following.

The two-dimensional form of Eq. (5.2) can be written as

$$\frac{\partial}{\partial t}(\rho\phi) + \frac{\partial J_x}{\partial x} + \frac{\partial J_y}{\partial y} = S,$$
(5.48)

where J_x and J_y are the total (convection plus diffusion) fluxes defined by

$$J_x \equiv \rho u \phi - \Gamma \frac{\partial \phi}{\partial x}$$
(5.49a)

and
$$J_y \equiv \rho v \phi - \Gamma \frac{\partial \phi}{\partial y},$$
(5.49b)

where u and v denote the velocity components in the x and y directions. The integration of Eq. (5.48) over the control volume shown in Fig. 5.9 would give

$$\frac{(\rho_P \phi_P - \rho_P^0 \phi_P^0) \, \Delta x \, \Delta y}{\Delta t} + J_e - J_w + J_n - J_s = (S_C + S_P \phi_P) \, \Delta x \, \Delta y,$$
(5.50)

where the source term has been linearized in the usual manner and, for the unsteady term, ρ_P and ϕ_P are assumed to prevail over the whole control volume. The "old" values (i.e., the values at the beginning of the time step)

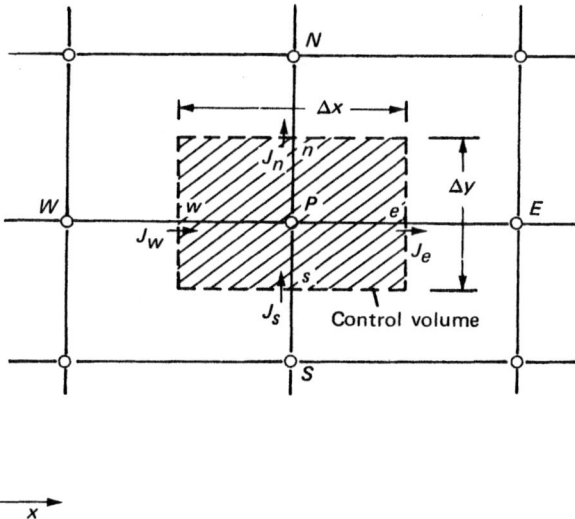

Figure 5.9 Control volume for the two-dimensional situation.

are denoted by ρ_P^0 and ϕ_P^0. In conformity with the fully implicit practice, all other values (i.e., those without a superscript) are to be regarded as the "new" values. The quantities J_e, J_w, J_n, and J_s are the *integrated* total fluxes over the control-volume faces; that is, J_e stands for $\int J_x\, dy$ over the interface e, and so on.

In a similar manner, we can integrate the continuity equation (5.1) over the control volume and obtain

$$\frac{(\rho_P - \rho_P^0)\, \Delta x\, \Delta y}{\Delta t} + F_e - F_w + F_n - F_s = 0 , \qquad (5.51)$$

where F_e, F_w, F_n, and F_s are the mass flow rates through the faces of the control volume. If ρu at point e is taken to prevail over the whole interface e, we can write

$$F_e = (\rho u)_e\, \Delta y . \qquad (5.52a)$$

Similarly,

$$F_w = (\rho u)_w\, \Delta y , \qquad (5.25b)$$

$$F_n = (\rho v)_n\, \Delta x , \qquad (5.52c)$$

$$F_s = (\rho v)_s\, \Delta x . \qquad (5.52d)$$

If we now multiply Eq. (5.51) by ϕ_P and substract it from Eq. (5.50), we obtain

$$(\phi_P - \phi_P^0)\, \frac{\rho_P^0\, \Delta x\, \Delta y}{\Delta t} + (J_e - F_e\phi_P) - (J_w - F_w\phi_P) + (J_n - F_n\phi_P)$$

$$- (J_s - F_s\phi_P) = (S_C + S_P\phi_P)\, \Delta x\, \Delta y . \qquad (5.53)$$

This manipulation of Eqs. (5.51) and (5.50) to obtain Eq. (5.53) is the discretization analogue of the combination of Eqs. (5.1) and (5.2) to derive Eq. (5.3). An alternative would have been to start the derivation of the discretization equation from Eq. (5.3); but this alternative is not as convenient.

The assumption of uniformity over a control-volume face enables us to employ our one-dimensional practices for the two-dimensional situation. At this point, we recall that Eqs. (5.44) and (5.45) provided a way of expressing terms such as $J_e - F_e\phi_P$ and $J_w - F_w\phi_P$. We use this here in the following manner:

$$J_e - F_e \phi_P = a_E(\phi_P - \phi_E) , \qquad (5.54a)$$

$$J_w - F_w \phi_P = a_W(\phi_W - \phi_P) , \qquad (5.54b)$$

where

$$a_E = D_e A(|P_e|) + [\![-F_e, 0]\!] , \qquad (5.55a)$$

$$a_W = D_w A(|P_w|) + [\![F_w, 0]\!] . \qquad (5.55b)$$

Here D_e and D_w, like their counterparts F_e and F_w, contain the area Δy of the faces e and w [see Eqs. (5.58) in Section 5.3-2]. With similar expressions for $J_n - F_n \phi_P$ and $J_s - F_s \phi_P$, we are in a position to write the final form of the discretization equation. Because of the nature of the expressions in Eqs. (5.54), the rule about the sum of the neighbor coefficients is readily satisfied.

When the given velocity and density fields *do* satisfy the continuity discretization equation, the foregoing derivation and a derivation based on Eq. (5.50) alone will yield identical discretization equations. However, when the given flow field *does not* satisfy the continuity equation, the two formulations give different equations and lead to different solutions. We prefer the formulation that satisfies our basic rule, for the reasons given in Chapter 3.

How could we encounter flow fields that do not satisfy continuity? The possibility arises because often the flow field is not really *given* but is iteratively calculated, just as the temperature-dependent conductivity is updated in a conduction problem. Before the final convergence is attained, the imperfect flow field at intermediate stages of iteration may not satisfy the continuity equation. It is for this reason that we have taken special care to satisfy Rule 4.

5.3-2 The Final Discretization Equation

The two-dimensional discretization equation can now be written as

$$a_P \phi_P = a_E \phi_E + a_W \phi_W + a_N \phi_N + a_S \phi_S + b , \qquad (5.56)$$

where

$$a_E = D_e A(|P_e|) + [\![-F_e, 0]\!] , \qquad (5.57a)$$

$$a_W = D_w A(|P_w|) + [\![F_w, 0]\!] , \qquad (5.57b)$$

$$a_N = D_n A(|P_n|) + [\![-F_n, 0]\!] , \qquad (5.57c)$$

$$a_S = D_s A(|P_s|) + [\![F_s, 0]\!] , \qquad (5.57d)$$

$$a_P^0 = \frac{\rho_P^0 \, \Delta x \, \Delta y}{\Delta t} \, , \tag{5.57e}$$

$$b = S_C \, \Delta x \, \Delta y + a_P^0 \phi_P^0 \, , \tag{5.57f}$$

$$a_P = a_E + a_W + a_N + a_S + a_P^0 - S_P \, \Delta x \, \Delta y \, . \tag{5.57g}$$

Here ϕ_P^0 and ρ_P^0 refer to the known values at time t, while all other values (ϕ_P, ϕ_E, ϕ_W, ϕ_N, ϕ_S, and so on) are the unknown values at time $t + \Delta t$. The flow rates F_e, F_w, F_n, and F_s have been defined in Eqs. (5.52). The corresponding conductances are defined by

$$D_e = \frac{\Gamma_e \, \Delta y}{(\delta x)_e} \, , \tag{5.58a}$$

$$D_w = \frac{\Gamma_w \, \Delta y}{(\delta x)_w} \, , \tag{5.58b}$$

$$D_n = \frac{\Gamma_n \, \Delta x}{(\delta y)_n} \, , \tag{5.58c}$$

$$D_s = \frac{\Gamma_s \, \Delta x}{(\delta y)_s} \, , \tag{5.58d}$$

and the Peclet numbers by

$$P_e = \frac{F_e}{D_e} \qquad P_w = \frac{F_w}{D_w} \qquad P_n = \frac{F_n}{D_n} \qquad P_s = \frac{F_s}{D_s} \, . \tag{5.59}$$

The function $A(|P|)$ can be selected from Table 5.2 for the desired scheme. The power-law scheme is recommended, for which

$$A(|P|) = [\![0, (1 - 0.1|P|)^5]\!] \, . \tag{5.60}$$

It can be appreciated that even at this stage the physical significance of the various coefficients in Eq. (5.56) is easy to understand. The neighbor coefficients a_E, a_W, a_N, and a_S represent the convection and diffusion influence at the four faces of the control volume, in terms of the flow rate F and the conductance D. The term $a_P^0 \phi_P^0$ is the known ϕ content of the control volume (at time t) divided by the time step. The remaining terms can be similarly interpreted.

5.4 DISCRETIZATION EQUATION FOR THREE DIMENSIONS

At last, we have arrived at our destination. We set out to write a discretization equation based on the general differential equation (5.2). Now, here it is in three dimensions (with T and B representing the "top" and "bottom" neighbors in the z direction):

$$a_P \phi_P = a_E \phi_E + a_W \phi_W + a_N \phi_N + a_S \phi_S + a_T \phi_T + a_B \phi_B + b , \quad (5.61)$$

where

$$a_E = D_e A(|P_e|) + [\![-F_e, 0]\!] , \quad (5.62a)$$

$$a_W = D_w A(|P_w|) + [\![F_w, 0]\!] , \quad (5.62b)$$

$$a_N = D_n A(|P_n|) + [\![-F_n, 0]\!] , \quad (5.62c)$$

$$a_S = D_s A(|P_s|) + [\![F_s, 0]\!] , \quad (5.62d)$$

$$a_T = D_t A(|P_t|) + [\![-F_t, 0]\!] , \quad (5.62e)$$

$$a_B = D_b A(|P_b|) + [\![F_b, 0]\!] , \quad (5.62f)$$

$$a_P^0 = \frac{\rho_P^0 \, \Delta x \, \Delta y \, \Delta z}{\Delta t} , \quad (5.62g)$$

$$b = S_C \, \Delta x \, \Delta y \, \Delta z + a_P^0 \phi_P^0 , \quad (5.62h)$$

$$a_P = a_E + a_W + a_N + a_S + a_T + a_B + a_P^0 - S_P \, \Delta x \, \Delta y \, \Delta z . \quad (5.62i)$$

The flow rates and conductances are defined as

$$F_e = (\rho u)_e \, \Delta y \, \Delta z \qquad D_e = \frac{\Gamma_e \, \Delta y \, \Delta z}{(\delta x)_e} , \quad (5.63a)$$

$$F_w = (\rho u)_w \, \Delta y \, \Delta z \qquad D_w = \frac{\Gamma_w \, \Delta y \, \Delta z}{(\delta x)_w} , \quad (5.63b)$$

$$F_n = (\rho v)_n \, \Delta z \, \Delta x \qquad D_n = \frac{\Gamma_n \, \Delta z \, \Delta x}{(\delta y)_n} , \quad (5.63c)$$

$$F_s = (\rho v)_s \, \Delta z \, \Delta x \qquad D_s = \frac{\Gamma_s \, \Delta z \, \Delta x}{(\delta y)_s} , \quad (5.63d)$$

$$F_t = (\rho w)_t \, \Delta x \, \Delta y \qquad D_t = \frac{\Gamma_t \, \Delta x \, \Delta y}{(\delta z)_t} \, , \qquad (5.63e)$$

$$F_b = (\rho w)_b \, \Delta x \, \Delta y \qquad D_b = \frac{\Gamma_b \, \Delta x \, \Delta y}{(\delta z)_b} \, . \qquad (5.63f)$$

The Peclet number P is to be taken as the ratio of F and D; thus, $P_e = F_e/D_e$, and so on. The function $A(|P|)$ is listed in Table 5.2 for various schemes. The power-law formulation is

$$A(|P|) = [\![0, (1 - 0.1|P|)^5]\!] \, . \qquad (5.64)$$

5.5 A ONE–WAY SPACE COORDINATE

In Chapter 2 we noted that coordinates can be classified as one-way and two-way, and that the identification of a one-way coordinate offers some computational advantages. Time is a one-way coordinate, and we have used it as such in formulating a marching procedure in time. The convection-diffusion formulation reveals that a space coordinate can also become one-way.

5.5-1 What Makes a Space Coordinate One-Way

We have seen from Fig. 5.4 or 5.6 that the coefficient of a downstream neighbor becomes small when the Peclet number is large. When the Peclet number exceeds 10, the power-law scheme will set the downstream-neighbor coefficient equal to zero. (The hybrid scheme does this for a Peclet number greater than 2.) Suppose that, in the two-dimensional situation shown in Fig. 5.10, there is a high flow rate in the positive x direction. Then, for all the grid points P along a y-direction line, the coefficients a_E will be zero. In other words, ϕ_P will be dependent on ϕ_W, ϕ_N, and ϕ_S, but not on ϕ_E. Thus, the x coordinate will become a one-way coordinate since the ϕ value at any point will be uninfluenced by any of the downstream values. A marching solution procedure would then be possible in the x direction.

Even when a space coordinate is not one-way over the whole calculation domain, its *local* one-way behavior is often useful in formulating the boundary conditions. This is discussed next.

5.5-2 The Outflow Boundary Condition

We described the treatment of the boundary conditions in some detail in Chapter 4. It has been tacitly assumed that the same treatment applies to the convection-diffusion problem. However, at an "outflow" boundary, i.e., where the fluid *leaves* the calculation domain, one normally knows neither the value

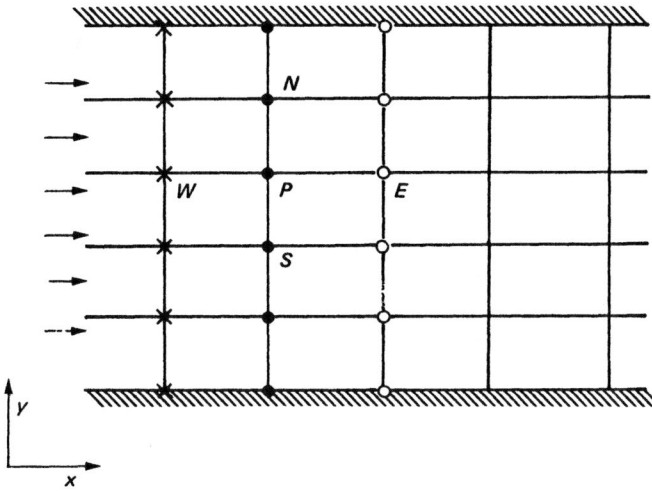

Figure 5.10 Situation with a one-way space coordinate.

of ϕ nor its flux. At the outflow boundary shown in Fig. 5.11, for example, one may not know the temperature or the heat flux. How can we then solve the problem? The answer is surprisingly simple: No boundary-condition information is needed at an outflow boundary. Consider the grid shown in the inset of Fig. 5.11. For all grid points P next to the outflow boundary, the coefficient a_E will be zero if the Peclet number is sufficiently large. Thus, the coefficients multiplying the boundary values will all be zero, and hence no boundary values will be needed. In other words, the region near the outflow boundary exhibits, for large Peclet numbers, local one-way behavior; since the

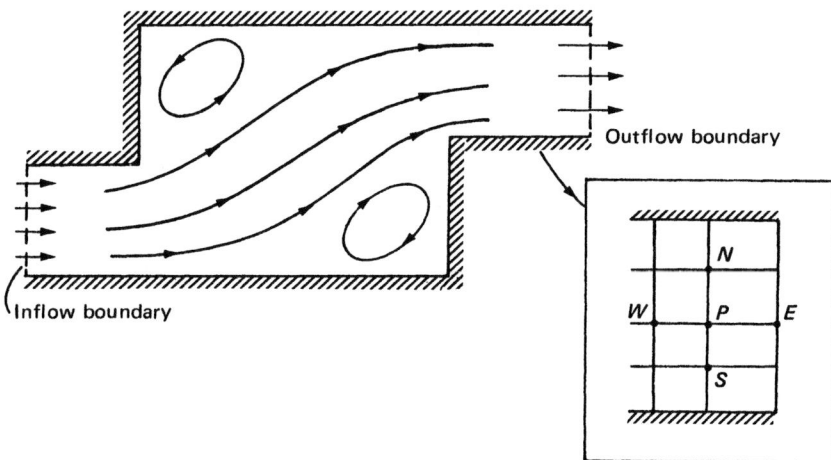

Figure 5.11 Example of the outflow boundary.

boundary points are *downstream* of the calculation domain, they do not influence the solution.

It is true that the above argument is based on the Peclet number being sufficiently large. But, in the absence of any other boundary-condition information, we can always assume the diffusion coefficient Γ at an outflow boundary to be small and thus work with a large Peclet number. An assumption such as this, which is a slight distortion of reality, is what we must resort to if we are to get meaningful solutions in the absence of any further information about the outflow boundary. The resulting inaccuracy, if there is any at all, is the price we pay for the freedom to isolate the calculation domain from the universe that lies downstream of the outflow boundary.

If the neglect of the diffusion at an outflow boundary appears, for some reason, to be serious, then we should conclude that the analyst has placed the outflow boundary at an inappropriate location. A repositioning of the boundary would normally make the outflow treatment acceptable. A particularly bad choice of an outflow-boundary location is the one in which there is an "inflow" over a part of it. An example of this is shown in Fig. 5.12. For such a bad choice of the boundary, no meaningful solution can be obtained.

This may be a convenient place to review the boundary-condition practices for convection-diffusion problems. Whenever there is no fluid flow across the boundary of the calculation domain, the boundary flux is purely a diffusion flux, and the practices described in Chapter 4 apply. For those parts of the boundary where the fluid flows *into* the domain, usually the values of ϕ are known. (The problem is not properly specified if we do not know the value of ϕ that a fluid stream brings with it.) The parts of the boundary where the fluid *leaves* the calculation domain form the outflow boundary, which we have already discussed.

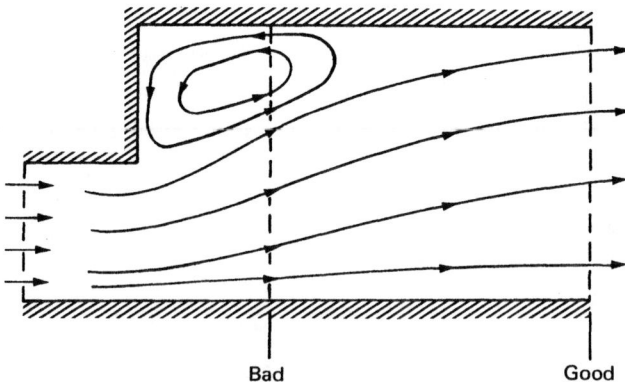

Figure 5.12 Good and bad choices of the location of the outflow boundary.

5.6 FALSE DIFFUSION

In this section, we shall discuss a topic that has caused considerable controversy, confusion, and misunderstanding among the practitioners of numerical analysis. There is something called "false diffusion," which is quite commonly misinterpreted, but which, in its proper meaning, represents a major weak point of most convection-diffusion formulations.

5.6-1 The Common View of False Diffusion

It is very common to encounter, in the literature, statements such as (1) the central-difference scheme has second-order accuracy, while the upwind scheme is only first-order accurate; or (2) the upwind scheme causes severe false diffusion. The implication is that the central-difference scheme is better than the upwind scheme.

It is true that from a Taylor-series expansion one can show that the central-difference scheme has a truncation error of the order of $(\Delta x)^2$, while the upwind scheme has an error of the order of (Δx). However, since the $\phi \sim x$ variation arising in the convection-diffusion problem is exponential, a truncated Taylor series ceases to be a good representation of it for anything but extremely small values of Δx (or, rather, of the corresponding Peclet number). At larger values of Δx, which is all one can afford in most practical problems, the Taylor-series analysis is misleading; there, as we have seen, it is the upwind scheme that gives more reasonable results than the central-difference scheme.

If we compare the coefficients for the central-difference and upwind schemes [Eqs. (5.11) and (5.15)], it can be shown that the upwind scheme is equivalent to replacing Γ in the central-difference scheme with $\Gamma + \rho u \delta x/2$. In other words, the upwind scheme seems to augment the true diffusion coefficient Γ by a fictitious (and hence false) diffusion coefficient $\rho u \delta x/2$. This introduction of an artificial diffusion coefficient is then considered to be inaccurate, a wrong representation of reality, and hence bad. Again, the trouble in the argument lies in assuming the central-difference scheme as accurate and standard (or the underlying Taylor-series expansion as reliable) and then viewing the upwind scheme from this frame of reference. In this manner, one would discover some false diffusion even in the exponential scheme, which is the exact solution itself. On the other hand, the theory presented in this chapter leads to the conclusion that the so-called false diffusion coefficient $\rho u \delta x/2$ is indeed a desirable addition at large Peclet numbers, for it actually tends to correct the wrong implications that would otherwise follow from the central-difference scheme.

There is no doubt that, for very small Peclet numbers, the central-difference scheme *is* more accurate than the upwind scheme. This has already been shown in a number of diagrams; and our favored schemes such as the exponential, the hybrid, and the power-law scheme indeed conform to the central-difference scheme at very low Peclet numbers. In any case, the question of false diffusion is never serious at low Peclet numbers, because then the real diffusion is quite large by comparison. It is for large Peclet numbers that the matter of false diffusion attains importance. There the central-difference scheme has little to offer, and all the other schemes that we have considered show almost identical behavior. It is for this reason that our remaining discussion will concentrate on very large Peclet numbers and on the upwind scheme; however, the conclusions will be equally applicable to the exponential, hybrid, and power-law schemes.

5.6-2 The Proper View of False Diffusion

Having seen that the common view of false diffusion is indeed misleading, we now turn to what can be truly described as false diffusion. The first thing to recognize is that false diffusion is a *multidimensional* phenomenon; it has absolutely no counterpart in steady one-dimensional situations. (*Unsteady* one-dimensional situations do suffer from a kind of false diffusion; we shall, however, confine our attention to steady situations.)

To visualize what is correctly meant by false diffusion, let us consider the situation shown in Fig. 5.13. Two parallel streams of equal velocity but unequal temperatures come in contact. If the diffusion coefficient Γ is nonzero, a mixing layer will form in which the temperature gradually changes from the higher value to the lower one, and the cross-stream width of this layer will grow in the downstream direction. If, on the other hand, the diffusion coefficient Γ were zero, no mixing layer would form and the temperature discontinuity would persist in the streamwise direction. The best situation for observing false diffusion is the one in which the real diffusion is set to zero. If the numerical solution for the $\Gamma = 0$ case produces a smeared temperature profile (which is characteristic of a nonzero Γ), we can conclude that the numerical scheme entails false diffusion.

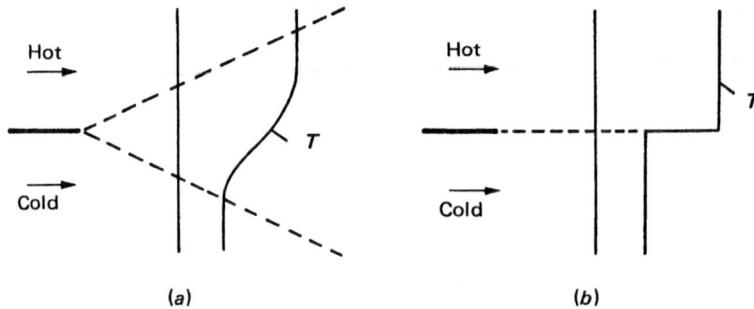

Figure 5.13 Temperature distributions in the presence and absence of diffusion. (*a*) $\Gamma \neq 0$; (*b*) $\Gamma = 0$.

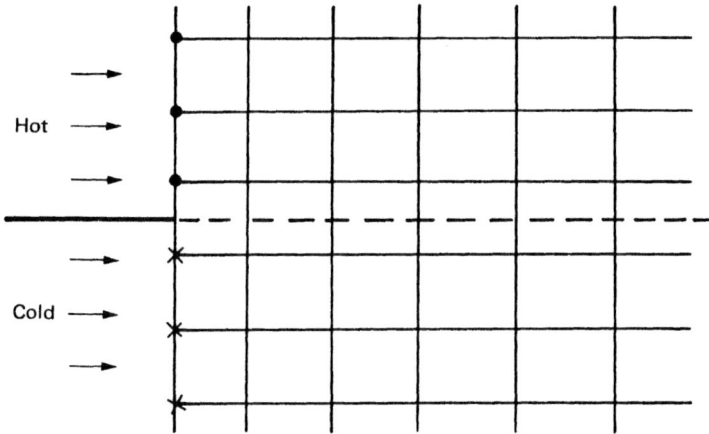

Figure 5.14 Situation with flow along the x direction.

For $\Gamma = 0$, the central-difference scheme would lead to $a_P = 0$. Therefore, the usual iterative methods for solving the algebraic equations cannot be used. If an attempt is made to solve the equations by a direct method, then either a unique solution is not found or the solutions turn out to be highly unrealistic.

Implications of the upwind scheme. We shall now try to solve the problem shown in Fig. 5.13b by the upwind scheme for two orientations of the grid.

1. *Uniform flow in the* x *direction.* Let us consider the situation shown in Fig. 5.14. The flow is aligned in the x direction, and the left-hand boundary has known temperatures with a sharp discontinuity. Since Γ is zero and there is no flow in the y direction, the coefficients a_N and a_S will be zero. The coefficient a_E of the *downstream* neighbor will also be zero. Thus, a_P must be equal to a_W, and this leads to

$$\phi_P = \phi_W . \tag{5.65}$$

As a result, the given upstream value on each horizontal line will become established at all points on that line. The temperature discontinuity in the upstream profile will then be preserved. No false diffusion is, therefore, encountered here.

2. *Uniform flow at 45° to the grid lines.* The situation changes greatly when the same problem is solved on a grid in which the grid lines are inclined at 45° to the flow direction. Let us, for convenience, use a uniform grid with $\Delta x = \Delta y$. The flow velocities in the x and y directions are equal. The result is that the coefficients of the upstream neighbors, a_W and a_S, become equal, while those of the downstream neighbors, a_E and a_N, turn out to be zero. Thus, we have

$$\phi_P = 0.5 \ \phi_W + 0.5 \ \phi_S \ . \tag{5.66}$$

For the grid shown in Fig. 5.15, the temperature discontinuity is represented by setting the left-boundary temperatures* equal to 100, and the bottom-boundary temperatures equal to zero. The resulting solution at the interior points is written adjacent to each grid point. If there were no false diffusion, we would have obtained a value of 100 above the diagonal through the lower left corner, and a value of zero below the diagonal. On the other hand, the actual solution obtained does represent a smeared temperature profile, much like the one in Fig. 5.13a.

   Remarks. (1) The false diffusion occurs when the flow is oblique to the grid lines and when there is a nonzero gradient of the dependent variable in the direction normal to the flow. (2) An approximate expression for the false diffusion coefficient for a two-dimensional situation has been given by de Vahl Davis and Mallinson (1972); it is

$$\Gamma_{\text{false}} = \frac{\rho U \ \Delta x \ \Delta y \ \sin \ 2\theta}{4(\Delta y \ \sin^3 \ \theta + \Delta x \ \cos^3 \ \theta)} \ , \tag{5.67}$$

where U is the resultant velocity, and θ is the angle (between 0 and 90°) made by the velocity vector with the x direction. It is easy to see from this equation that no false diffusion is present when the resultant flow is along one of the sets of grid lines; on the other hand, the false diffusion is most serious when the flow direction makes an angle of 45° with the grid lines. (3) The amount of false diffusion can be reduced by using smaller Δx and Δy and, whenever possible, by orienting the grid such that the grid lines more or less align with the flow direction. (4) Since real diffusion is present in many problems, it is then sufficient to make the false diffusion small in comparison with the real diffusion. (5) The use of the central-difference scheme is no remedy for false diffusion. As mentioned earlier, the central-difference scheme gives highly unrealistic solutions when large Peclet numbers are involved. (6) The basic cause of false diffusion is the practice of treating the flow across each control-volume face as locally one-dimensional. For the situation shown in the inset of Fig. 5.15, the value of ϕ convected by the oblique flow to the grid point P actually comes from the corner grid point SW. However, this convection is represented as the effect of two separate streams coming from the grid points W and S. (7) Schemes that would give less false diffusion should take account of the multidimensional nature of the flow. It would also

---

*It may appear that the temperatures along the left and bottom boundaries of the grid in Fig. 5.15 are not really known from the problem specification of Fig. 5.13b. However, once the exact solution for a problem is known, any domain over which the exact solution is valid can be chosen, and the boundary values can be prescribed from the exact solution. This method of constructing test problems that have known exact solutions has been used by Runchal (1972).

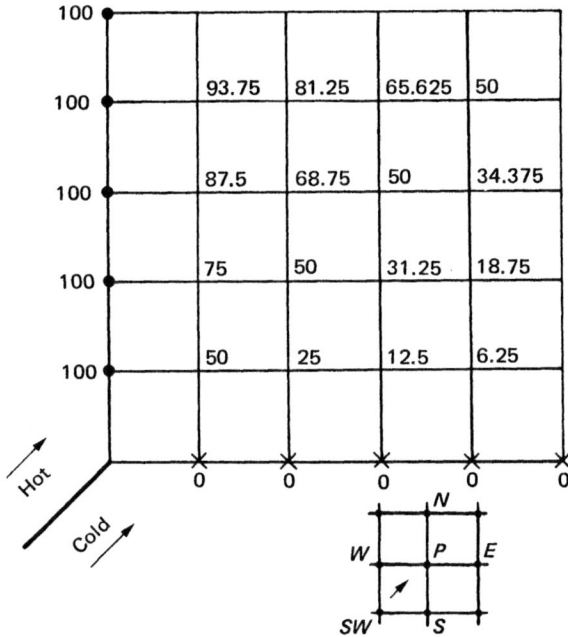

Figure 5.15 Situation with flow at 45° to the grid lines.

be necessary to involve more neighbors in the discretization equation. Although a few such schemes have been worked out [for example, Raithby (1976b)] and have shown an impressive reduction in false diffusion, they are significantly more complicated and so far insufficiently tested. For these reasons, we shall not discuss them here. (8) A more detailed discussion of false diffusion has been given by Raithby (1976a).

5.7 CLOSURE

In this chapter, we have completed the construction of the general discretization equation for the dependent variable ϕ. The convection term was the only addition that we made here, but it led to a number of interesting considerations. Our formulation ensures physically realistic behavior and thus holds the key to successful computation in the presence of fluid flow. The flow field itself, of course, must also be calculated in most cases. It is to this matter that we turn our attention in the next chapter.

PROBLEMS

5.1 In a steady two-dimensional situation, the variable ϕ is governed by

$$\text{div} (\rho \mathbf{u} \phi) = \text{div} (\Gamma \text{ grad } \phi) + a - b\phi ,$$

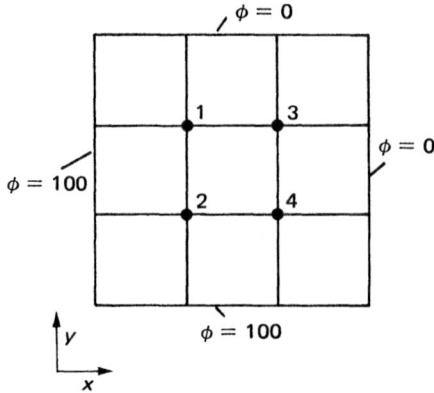

Figure 5.16 Boundary conditions for Problem 5.1.

where $\rho = 1$, $\Gamma = 1$, $a = 10$, and $b = 2$. The flow field is such that $u = 1$ and $v = 4$ everywhere. For the uniform grid shown in Fig. 5.16, $\Delta x = \Delta y = 1$. The values of ϕ are given for the four boundaries. Adopting the control-volume design according to Practice A in Section 4.6-1, calculate the values of ϕ_1, ϕ_2, ϕ_3, and ϕ_4 by use of:

(a) The central-difference scheme
(b) The upwind scheme
(c) The hybrid scheme
(d) The power-law scheme

5.2 Obtain the exact solution of the equation

$$\frac{d}{dx}\left(\rho u \phi - \Gamma \frac{d\phi}{dx}\right) = S,$$

where ρu, Γ, and S are all constant; the boundary conditions are $\phi = \phi_0$ at $x = 0$, and $\phi = \phi_L$ at $x = L$. Use the exponential scheme to obtain a numerical solution of the problem for various values of $\rho u L / \Gamma$ and $(SL^2/\Gamma)/(\phi_L - \phi_0)$. Do you get perfect agreement with the exact solution? Why?

5.3 A parallel-flow heat exchanger is governed by

$$m_h c_h \frac{dT_h}{dx} = \frac{UA}{L}(T_c - T_h) \quad \text{and} \quad m_c c_c \frac{dT_c}{dx} = \frac{UA}{L}(T_h - T_c),$$

where m, c, and T stand for the mass flow rate, the specific heat, and the temperature, respectively; the subscripts h and c denote the hot and cold fluids, respectively; U is the overall heat transfer coefficient between the two fluids; A is the total heat transfer area; and L is the length of the heat exchanger. The inlet temperatures $T_{h,in}$ and $T_{c,in}$ are given. Obtain a numerical solution for the dimensionless temperatures $(T_h - T_{c,in})/\Delta T$ and $(T_c - T_{c,in})/\Delta T$ as functions of x/L for the conditions $m_h c_h = m_c c_c$ and $UA/m_h c_h = 1$. The temperature difference ΔT equals $T_{h,in} - T_{c,in}$. Compare the numerical results with the exact solution. (Although the two coupled equations can be handled iteratively by sequential solution for T_h and T_c, a direct simultaneous solution is often advantageous for such a case. This can be achieved by use of the algorithm for two coupled variables, which was outlined in Problem 4.17.)

5.4 Consider the one-dimensional distribution of a variable ϕ governed by convection and diffusion. The flow field is created by the flow in a porous-walled duct; m_x denotes the x-direction mass flow rate along the duct at any location x, and m_L is the rate of mass leakage through the porous walls per unit length of the duct. Obviously, $dm_x/dx = -m_L$. The variable ϕ is governed by

$$\frac{d}{dx}(m_x\phi) + m_L\phi_L - \Gamma A \frac{d^2\phi}{dx^2} = 0 \, ,$$

where A is the duct cross section. When m_L is positive (i.e., fluid is leaking out), ϕ_L is to be taken as ϕ within the duct; when m_L is negative (i.e., fluid is leaking into the duct), ϕ_L is to be taken as ϕ_0, which is the value of ϕ in the ambient outside the duct. For a duct length of l, the boundary values are $\phi = \phi_1$ at $x = 0$ and $\phi = \phi_0$ at $x = l$. Assume m_L and ΓA to be constant. Use the central-difference and power-law schemes to find the dimensionless distribution of ϕ for the following two cases:

(a) At $x = 0$, $m_x l/\Gamma A = 40$; at $x = l$, $m_x = 0$
(b) At $x = 0$, $m_x = 0$; at $x = l$, $m_x l/\Gamma A = 40$.

5.5 Write Eq. (5.4) by replacing x with η, where η is defined as

$$\eta \equiv \int_0^x \frac{dx}{\Gamma} \, .$$

Hence show that, just as Eq. (5.17) is the solution of Eq. (5.4) for the case of uniform Γ, the solution for nonuniform Γ is given by

$$\frac{\phi - \phi_0}{\phi_L - \phi_0} = \frac{\exp(\rho u \eta) - 1}{\exp(\rho u \eta_L) - 1} \, ,$$

where η_L is the value of η at $x = L$. Note that $\rho u \eta_L$ is the Peclet number. If the derivation on these lines is continued, we get Eq. (5.22), where P_e must be defined as $P_e = (\rho u)_e(\delta\eta)_e$. Assuming that a grid-point value of Γ prevails throughout the control volume surrounding it, we can express $(\delta\eta)_e$ in terms of the Γ's and the distance increments (shown in Fig. 4.1). Hence, we have

$$P_e = (\rho u)_e \left[\frac{(\delta x)_{e-}}{\Gamma_P} + \frac{(\delta x)_{e+}}{\Gamma_E} \right] .$$

CALCULATION OF THE FLOW FIELD

6.1 NEED FOR A SPECIAL PROCEDURE

6.1-1 The Main Difficulty

In Chapter 5, we formulated the procedure for solving the general differential equation for ϕ in the presence of a *given* flow field. However, except in some very special circumstances, it is not possible to *specify* the flow field; rather, we must calculate the local velocity components and the density field from the appropriate governing equations. The velocity components are governed by the momentum equations, which are particular cases of the general differential equation for ϕ (with $\phi = u$, $\Gamma = \mu$, and so on). Thus, we are tempted to conclude that we already have developed the method for solving the momentum equations, thereby getting the velocity field. Where, then, is the difficulty?

If the nonlinearity of the momentum equations appears to be a difficulty, we only have to remind ourselves that, while treating heat conduction, we saw how to handle nonlinearity by iteration. In particular, the convection coefficient ρu being a function of the dependent variable u of the momentum equation is no different from the conductivity k being a function of the temperature T. Starting with a guessed velocity field, we could iteratively solve the momentum equations to arrive at the converged solution for the velocity components.

The real difficulty in the calculation of the velocity field lies in the unknown pressure field. The pressure gradient forms a part of the source term for a momentum equation. Yet, there is no obvious equation for obtaining pressure. For a *given* pressure field, it is true, there is no particular difficulty in solving the momentum equations. But, the way to determine the pressure field seems rather obscure.

The pressure field is indirectly specified via the continuity equation. When the *correct* pressure field is substituted into the momentum equations, the resulting velocity field satisfies the continuity equation. This indirect specification, however, is not very useful for our purposes unless we attempt a direct solution of the whole set of the discretization equations resulting from the momentum and continuity equations. Since we have preferred iterative methods of solving the discretization equations even for a single dependent variable, the direct solution for the entire set of velocity components and pressure seems out of the question.*

6.1-2 Vorticity-based Methods

The difficulty associated with the determination of pressure has led to methods that eliminate pressure from the governing equations. Thus, in two dimensions, the elimination of pressure from the two momentum equations by cross differentiation leads to a vorticity-transport equation. (This derivation is outlined in Problem 6.1.) This, when combined with the definition of a stream function for steady two-dimensional situations, is the basis of the well-known "stream-function/vorticity method" described by, among others, Dix (1963), Fromm and Harlow (1963), Pearson (1965), Barakat and Clark (1966), and Runchal and Wolfshtein (1969) and made easily accessible through the book by Gosman, Pun, Runchal, Spalding, and Wolfshtein (1969).

The stream-function/vorticity method has some attractive features. The pressure makes no appearance, and, instead of dealing with the continuity equation and two momentum equations, we need to solve only two equations to obtain the stream function and the vorticity. Some of the boundary conditions can be rather easily specified: When an external irrotational flow lies adjacent to the calculation domain, the boundary vorticity can conveniently be set equal to zero. There are, however, some major disadvantages to the stream-function/vorticity method. The value of vorticity at a wall is difficult to specify and is often the cause of trouble in getting a converged solution. The pressure, which has been so cleverly eliminated, frequently happens to be an important desired result or even an intermediate outcome required for the calculation of density and other fluid properties. Then, the

---

*Some methods, especially those dealing with compressible flows, regard the density ρ as the dependent variable of the continuity equation and then extract the pressure from it via an equation of state. This approach is, however, inapplicable to constant-density or incompressible flows. In such situations, it is the effect of pressure on velocity, and *not* on density, that is of primary importance.

effort of extracting pressure from vorticity offsets the computational savings obtained otherwise. But, above all, the major shortcoming of the method is that it cannot easily be extended to three-dimensional situations, for which a stream function does not exist. Since most practical problems are three-dimensional, a method that is intrinsically restricted to two dimensions suffers from a serious limitation.

For three dimensions, an approach based on vorticity uses six dependent variables, namely, the three components of the vorticity vector and the three components of the velocity-potential vector [see Aziz and Hellums (1967), for example]. Thus, the complexity is actually greater than that of treating the three velocity components and pressure directly. Also, the vorticity vector and the velocity-potential vector involve concepts that are harder to visualize and interpret than the meanings of the velocity components and pressure. In keeping with our desire to formulate physically meaningful and illuminating approaches, we seek a method that uses the so-called primitive variables, namely the velocity components and pressure.

Thus the main task in this chapter is to convert the indirect information in the continuity equation into a direct algorithm for the calculation of pressure. A few minor difficulties arise, which we shall discuss before we begin this task.

6.2 SOME RELATED DIFFICULTIES

6.2-1 Representation of the Pressure-Gradient Term

If we begin to construct the discretization form of the x-direction momentum equation for the one-dimensional situation shown in Fig. 6.1, the only new feature is the representation of the term $-dp/dx$ integrated over the control volume. The resulting contribution to the discretization equation is the pressure drop $p_w - p_e$, which is the net pressure force exerted on the control volume of unit cross-sectional area. To express $p_w - p_e$ in terms of the grid-point pressures, we may assume a piecewise-linear profile for pressure. Further, if the control-volume faces e and w are chosen to lie midway[*] between the respective grid points, we have

$$p_w - p_e = \frac{p_W + p_P}{2} - \frac{p_P + p_E}{2} = \frac{p_W - p_E}{2} . \tag{6.1}$$

[*]This assumption is made here only for algebraic convenience. When the control-volume faces are *not* midway, the difficulties being discussed here do not go away, but appear in a less clear form. Thus, the assumption of *midway* faces is not a cause of the difficulties, but makes the discussion easy to follow.

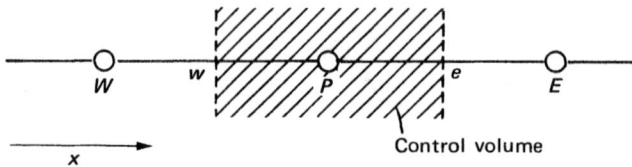

Figure 6.1 Three-grid-point cluster.

This means that the momentum equation will contain the pressure difference between two *alternate* grid points, and not between *adjacent* ones. The implication is that the pressure is, in effect, taken from a coarser grid than the one actually employed. This would tend to diminish the accuracy of the solution. But, there is another implication that is far more serious. It can be best seen from Fig. 6.2, where a pressure field is proposed in terms of the grid-point values of pressure. Such a zig-zag field cannot be regarded as realistic; but, for any grid point P, the corresponding $p_W - p_E$ can be seen to be zero, since the alternate pressure values are everywhere equal. Thus, the devastating consequence is that such a wavy pressure field will be *felt* like a uniform pressure field by the momentum equation.

The difficulty can be seen more dramatically in a two-dimensional situation. Just as the x-direction momentum is influenced by $p_W - p_E$, the y-direction momentum is affected by $p_S - p_N$; then the pressure p_P has no role to play. With this in mind, we can conclude that the pressure field shown in Fig. 6.3, which is made up of four arbitrary values of pressure arranged in a

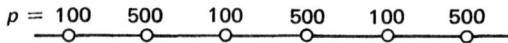

$p =$ 100   500   100   500   100   500

Figure 6.2 Zigzag pressure field.

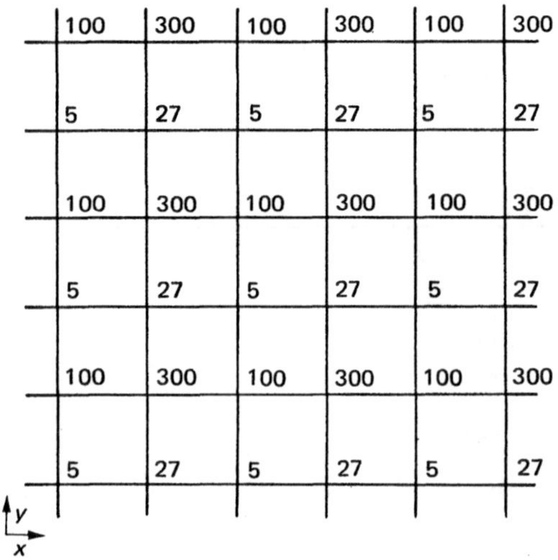

Figure 6.3 Checkerboard pressure field.

checkerboard pattern, would produce no pressure force in the x or y direction. Thus, a highly nonuniform pressure field would be treated as a uniform pressure field by the particular discretized form of the momentum equations. Should such pressure fields arise during the iterative solution procedure, there would be nothing to stop them from being preserved till convergence, since the momentum equations would be oblivious to their presence.

It should be noted that the actual numbers used in Figs. 6.2 and 6.3 do not have any particular significance; they simply indicate a pattern that can be constructed from any arbitrary numbers. It is easy to imagine that the three-dimensional situation would allow an even more complex pattern, which the momentum equations would still interpret as a uniform pressure field.

If a certain smooth pressure field is obtained as a solution, any number of additional solutions can be constructed by adding a checkerboard pressure field to that solution. The momentum equations would remain unaffected by this addition, since the checkerboard field implies zero pressure force. A numerical method that allows such absurd solutions is certainly undesirable.

6.2-2 Representation of the Continuity Equation

A similar kind of difficulty arises when we try to construct the discretization form of the continuity equation. For the steady one-dimensional constant-density situation, the continuity equation is simply

$$\frac{du}{dx} = 0 . \tag{6.2}$$

If we integrate this over the control volume shown in Fig. 6.1, we have

$$u_e - u_w = 0 . \tag{6.3}$$

Once again, the use of a piecewise-linear profile for u and of the *midway* locations of the control-volume faces leads to

$$\frac{u_P + u_E}{2} - \frac{u_W + u_P}{2} = 0 \tag{6.4}$$

or

$$u_E - u_W = 0 . \tag{6.5}$$

Thus, the discretized continuity equation demands the equality of velocities at *alternate* grid points and not at adjacent ones. A consequence is that velocity fields of the type shown in Fig. 6.4, which are not at all realistic, do satisfy the discretized continuity equation (6.5). In two- and three-dimensional situations, similar patterns for all the velocity components can be created;

$u =$ 100   400   100   400   100   400

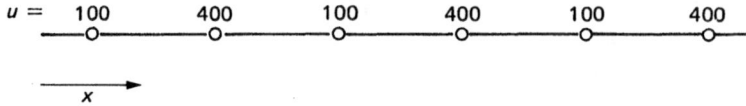

Figure 6.4 Wavy velocity field.

they will satisfy the continuity equation but can hardly be accepted as reasonable or meaningful solutions.

These difficulties must be resolved before a numerical method involving the velocity components and pressure can be formulated. In the literature, some methods can be found that pay no special attention to these difficulties. There, the possible unrealistic solutions are avoided by some special treatment at the boundaries, by overspecification of the boundary conditions, by underrelaxation with respect to a smooth initial guess, or by good fortune. But most such methods would accept pressure and velocity fields of the type shown in Figs. 6.2–6.4 as satisfactory solutions, and, in absence of special tricks, there is always the danger of arriving at such solutions.

Before we proceed to describe a way out of these difficulties, it is interesting to note that the troublesome hurdles in numerical analysis seem to be associated with the first derivatives. The second derivative is always well behaved and creates no difficulties. On the other hand, all the complications encountered in Chapter 5 can be attributed to the first derivative representing the convection term; and here, the first derivatives of pressure (in the momentum equations) and of velocity (in the continuity equation) cause considerable nuisance.

6.3 A REMEDY: THE STAGGERED GRID

The difficulties described so far can be resolved by recognizing that we do not have to calculate all the variables for the same grid points. We can, if we wish, employ a different grid for each dependent variable. Of course, we would not exercise this freedom if there were no benefit to be derived. But, in the case of the velocity components, there is a significant benefit to be obtained by arranging them on grids that are different from the grid used for all other variables. The benefit is that the difficulties described in Section 6.2 will totally disappear.

Such a displaced or "staggered" grid for the velocity components was first used by Harlow and Welch (1965) in their MAC method and has been used in other methods developed by Harlow and co-workers. It forms the basis of the SIVA procedure of Caretto, Curr, and Spalding (1972) and the SIMPLE procedure of Patankar and Spalding (1972a).

In the staggered grid, the velocity components are calculated for the points that lie on the faces of the control volumes. Thus, the x-direction

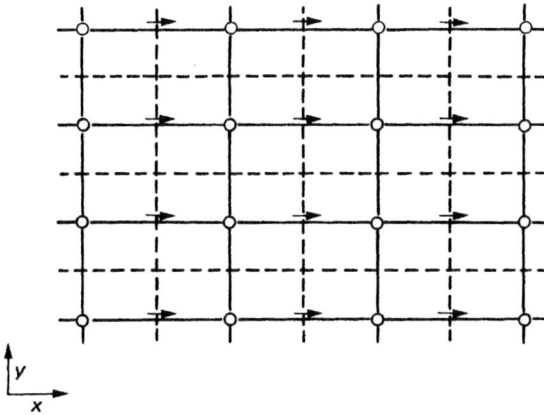

Figure 6.5 Staggered locations for u.

velocity u is calculated at the faces that are normal to the x direction. The locations for u are shown in Fig. 6.5 by short arrows, while the grid points (hereafter called the *main* grid points) are shown by small circles; the dashed lines indicate the control-volume faces. It will be noticed that, with respect to the main grid points, the u locations are staggered only in the x direction. In other words, the location for u lies on the x-direction link joining two adjacent main grid points. Whether the u location is exactly midway between the grid points depends upon how the control volumes are defined. The u location must lie on the control-volume face, irrespective of whether the latter happens to be midway between the grid points.

It is easy to see how the locations for the velocity components v and w are to be defined. In Fig. 6.6, a two-dimensional grid pattern is shown, with

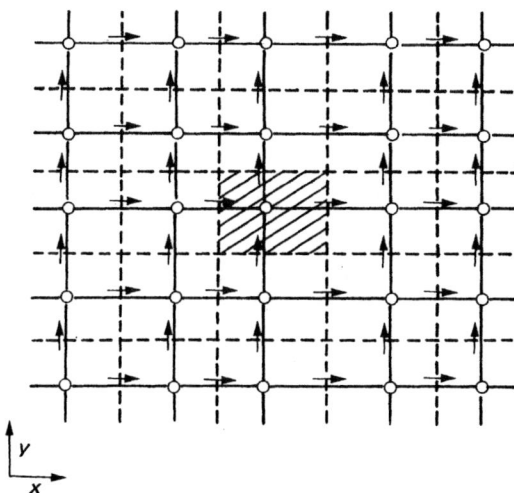

Figure 6.6 Staggered locations for u and v. $\rightarrow = u$; $\uparrow = v$; $\circ =$ other variables.

the locations for u and v placed on the respective control-volume faces. A corresponding three-dimensional pattern can be imagined in a straightforward manner.

An immediate consequence of the staggered grid is that the mass flow rates across the control-volume faces (the F's encountered in Chapter 5) can be calculated without any interpolation for the relevant velocity component. However, this feature, although it offers some convenience in setting up the general discretization equation for ϕ, is not an important advantage of the staggered grid.

The important advantages are twofold. For a typical control volume (shown shaded in Fig. 6.6) it is easy to see that the discretized continuity equation would contain the differences of *adjacent* velocity components, and that this would prevent a wavy velocity field, such as the one in Fig. 6.4, from satisfying the continuity equation. In the staggered grid, only "reasonable" velocity fields would have the possibility of being acceptable to the continuity equation. The second important advantage of the staggered grid is that the pressure difference between two *adjacent* grid points now becomes the natural driving force for the velocity component located between these grid points. Consequently, pressure fields such as those in Figs. 6.2 and 6.3 would no longer be felt as uniform pressure fields and could not arise as possible solutions.

The difficulties described in Section 6.2 can thus be attributed to the practice of calculating all variables for the same grid points; with the staggered grid, these difficulties are entirely eliminated.

This freedom from difficulties has its own price. A computer program based on the staggered grid must carry all the indexing and geometric information about the locations of the velocity components and must perform certain rather tiresome interpolations. But the benefits of the staggered grid are well worth the additional trouble.

6.4 THE MOMENTUM EQUATIONS

We again remind the reader that, if the pressure field is given, the solution of the momentum equations can be obtained by employing the formulation completed in Chapter 5 for the general variable ϕ. In the momentum equation, ϕ stands for the relevant velocity component, and Γ and S are to be given their appropriate meanings. The adoption of the staggered grid does make the discretized momentum equations somewhat different from the discretization equations for the other ϕ's that are calculated for the main grid points. But this difference is one of detail and not of essence. It arises from the use of staggered control volumes for the momentum equations.

A staggered control volume for the x-momentum equation is shown in Fig. 6.7. If we focus attention on the locations for u only, there is nothing

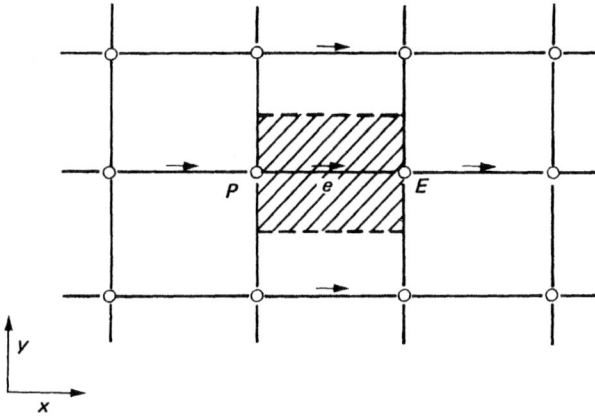

Figure 6.7 Control volume for u.

unusual about this control volume. Its faces lie between the point e and the corresponding locations for the neighbor u's. The control volume is, however, staggered in relation to the normal control volume around the main grid point P. The staggering is in the x direction only, such that the faces normal to that direction pass through the main grid points P and E. This layout realizes one of the main advantages of the staggered grid: The difference $p_P - p_E$ can be used to calculate the pressure force acting on the control volume for the velocity u.

The calculation of the diffusion coefficient and the mass flow rate at the faces of the u control volume shown in Fig. 6.7 would require an appropriate interpolation; but essentially the same formulation as described in Chapter 5 would be applicable. The resulting discretization equation can be written as

$$a_e u_e = \Sigma\, a_{nb} u_{nb} + b + (p_P - p_E)A_e . \qquad (6.6)$$

Here the number of neighbor terms will depend on the dimensionality of the problem. For the two-dimensional situation in Fig. 6.7, four u neighbors are shown outside the control volume; for a three-dimensional case, six neighbor u's would be included. The neighbor coefficients a_{nb} account for the combined convection-diffusion influence at the control-volume faces. The term b is defined in the same manner as in Eq. (5.57) or (5.62), *but the pressure gradient is not included in the source-term quantities* S_C *and* S_P. The pressure gradient gives rise to the last term in Eq. (6.6). Since the pressure field is also to be ultimately calculated, it would be inconvenient to bury the pressures in the momentum source term. The term $(p_P - p_E)A_e$ is the pressure force acting on the u control volume, A_e being the area on which the pressure difference acts. For two dimensions, A_e will be $\Delta y \times 1$, while in the three-dimensional case A_e will stand for $\Delta y\, \Delta z$.

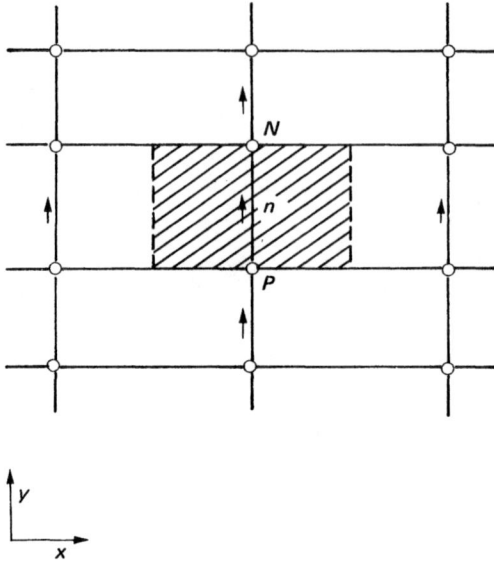

Figure 6.8 Control volume for v.

The momentum equations for the other directions are handled in a similar manner. Figure 6.8 shows the control volume for the y-direction momentum equation; it is staggered in the y direction. The discretization equation for v_n can be seen to be

$$a_n v_n = \Sigma\, a_{\text{nb}} v_{\text{nb}} + b + (p_P - p_N)A_n \,, \qquad (6.7)$$

where $(p_P - p_N)A_n$ is the appropriate pressure force. For the three-dimensional case, a similar equation for the velocity component w can be written.

The momentum equations can be solved only when the pressure field is given or is somehow estimated. Unless the correct pressure field is employed, the resulting velocity field will not satisfy the continuity equation. Such an imperfect velocity field based on a guessed pressure field p^* will be denoted by u^*, v^*, w^*. This "starred" velocity field will result from the solution of the following discretization equations:

$$a_e u_e^* = \Sigma\, a_{\text{nb}} u_{\text{nb}}^* + b + (p_P^* - p_E^*)A_e \,, \qquad (6.8)$$

$$a_n v_n^* = \Sigma\, a_{\text{nb}} v_{\text{nb}}^* + b + (p_P^* - p_N^*)A_n \,, \qquad (6.9)$$

$$a_t w_t^* = \Sigma\, a_{\text{nb}} w_{\text{nb}}^* + b + (p_P^* - p_T^*)A_t \,. \qquad (6.10)$$

In these equations, the velocity components and pressure have been given the superscript *. The location t, it can be noted, lies on the z-direction grid line between the grid points P and T.

6.5 THE PRESSURE AND VELOCITY CORRECTIONS

Our aim is to find a way of improving the guessed pressure p^* such that the resulting starred velocity field will progressively get closer to satisfying the continuity equation. Let us propose that the correct pressure p is obtained from

$$p = p^* + p' , \tag{6.11}$$

where p' will be called the *pressure correction*. Next, we need to know how the velocity components respond to this change in pressure. The corresponding velocity corrections u', v', w' can be introduced in a similar manner:

$$u = u^* + u' \qquad v = v^* + v' \qquad w = w^* + w' . \tag{6.12}$$

If we subtract Eq. (6.8) from Eq. (6.6), we have

$$a_e u'_e = \Sigma\, a_{nb} u'_{nb} + (p'_P - p'_E)A_e . \tag{6.13}$$

At this point, we shall boldly decide to drop the term $\Sigma\, a_{nb} u'_{nb}$ from the equation. An extensive discussion of this action will be presented in Section 6.7-2. For the time being, it is best to pay no attention to this move or to regard it simply as a computational convenience. The result is

$$a_e u'_e = (p'_P - p'_E)A_e \tag{6.14}$$

or
$$u'_e = d_e (p'_P - p'_E) , \tag{6.15}$$

where

$$d_e \equiv \frac{A_e}{a_e} . \tag{6.16}$$

Equation (6.15) will be called the velocity-correction formula, which can also be written as

$$u_e = u_e^* + d_e (p'_P - p'_E) . \tag{6.17}$$

This shows how the starred velocity u_e^* is to be corrected in response to the pressure corrections to produce u_e.

The correction formulas for the velocity components in other directions can be written similarly:

$$v_n = v_n^* + d_n(p_P' - p_N'),$$ (6.18)

$$w_t = w_t^* + d_t(p_P' - p_T').$$ (6.19)

Thus, we now have all the preparation needed for obtaining a discretization equation for p'. It is to this task that we now turn.

6.6 THE PRESSURE–CORRECTION EQUATION

We shall now turn the continuity equation into an equation for the pressure correction. For the purpose of this derivation, we shall assume that the density ρ does not directly depend on pressure. Later, the implications of this assumption will be discussed. The derivation is given here for the three-dimensional situation; the one- and two-dimensional forms can easily be obtained.

The continuity equation is

$$\frac{\partial \rho}{\partial t} + \frac{\partial(\rho u)}{\partial x} + \frac{\partial(\rho v)}{\partial y} + \frac{\partial(\rho w)}{\partial z} = 0.$$ (6.20)

We shall integrate this over the shaded control volume shown in Fig. 6.9. (Only a two-dimensional view is shown for convenience.) The same control volume, it will be remembered, was used for deriving the discretization equation for the general variable ϕ. For the integration of the term $\partial \rho / \partial t$, we

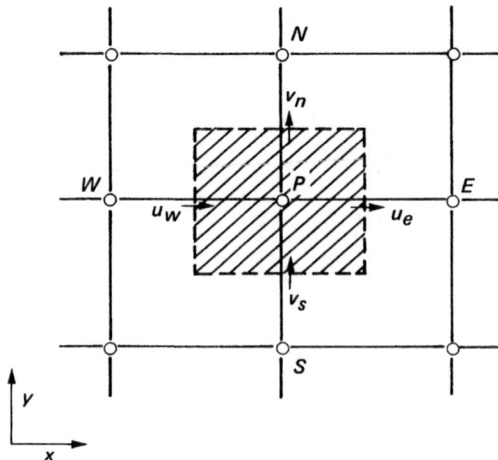

Figure 6.9 Control volume for the continuity equation.

shall assume that the density ρ_P prevails over the control volume. Also, a velocity component such as u_e located on a control-volume face will be supposed to govern the mass flow rate for the whole face. In conformity with the fully implicit practice, the *new* values of velocity and density (i.e., those at time $t + \Delta t$) will be assumed to prevail over the time step; the *old* density ρ_P^0 (i.e., the one at time t) will appear only through the term $\partial\rho/\partial t$.

With these decisions, the integrated form of Eq. (6.20) becomes

$$\frac{(\rho_P - \rho_P^0)\,\Delta x\,\Delta y\,\Delta z}{\Delta t} + [(\rho u)_e - (\rho u)_w]\,\Delta y\,\Delta z$$

$$+ [(\rho v)_n - (\rho v)_s]\,\Delta z\,\Delta x + [(\rho w)_t - (\rho w)_b]\,\Delta x\,\Delta y = 0 . \quad (6.21)$$

If we now substitute for all the velocity components the expressions given by the velocity-correction formulas [such as Eqs. (6.17)–(6.19)], we obtain, after rearrangement, the following discretization equation for p':

$$a_P p'_P = a_E p'_E + a_W p'_W + a_N p'_N + a_S p'_S + a_T p'_T + a_B p'_B + b , \quad (6.22)$$

where

$$a_E = \rho_e d_e\,\Delta y\,\Delta z , \quad\quad\quad\quad\quad\quad\quad\quad\quad\quad\quad\quad\quad (6.23a)$$

$$a_W = \rho_w d_w\,\Delta y\,\Delta z , \quad\quad\quad\quad\quad\quad\quad\quad\quad\quad\quad\quad\quad (6.23b)$$

$$a_N = \rho_n d_n\,\Delta z\,\Delta x , \quad\quad\quad\quad\quad\quad\quad\quad\quad\quad\quad\quad\quad (6.23c)$$

$$a_S = \rho_s d_s\,\Delta z\,\Delta x , \quad\quad\quad\quad\quad\quad\quad\quad\quad\quad\quad\quad\quad (6.23d)$$

$$a_T = \rho_t d_t\,\Delta x\,\Delta y , \quad\quad\quad\quad\quad\quad\quad\quad\quad\quad\quad\quad\quad (6.23e)$$

$$a_B = \rho_b d_b\,\Delta x\,\Delta y , \quad\quad\quad\quad\quad\quad\quad\quad\quad\quad\quad\quad\quad (6.23f)$$

$$a_P = a_E + a_W + a_N + a_S + a_T + a_B , \quad\quad\quad\quad\quad\quad\quad (6.23g)$$

$$b = \frac{(\rho_P^0 - \rho_P)\,\Delta x\,\Delta y\,\Delta z}{\Delta t} + [(\rho u^*)_w - (\rho u^*)_e]\,\Delta y\,\Delta z$$

$$+ [(\rho v^*)_s - (\rho v^*)_n]\,\Delta z\,\Delta x + [(\rho w^*)_b - (\rho w^*)_t]\,\Delta x\,\Delta y \quad (6.23h)$$

Since the values of the density ρ will normally be available only at the main grid points, the interface densities such as ρ_e may be calculated by any convenient interpolation. Whatever the method of interpolation, the value of ρ_e must be consistently used for the two control volumes to which the interface belongs (see basic Rule 1 in Chapter 3).

It can be seen from Eq. (6.23h) that the term b in the pressure-correction equation is essentially (the negative of) the left-hand side of the discretized continuity equation (6.21) evaluated in terms of the starred velocities. If b is zero, it means that the starred velocities, in conjunction with the available value of $(\rho_P^0 - \rho_P)$, do satisfy the continuity equation, and no pressure correction is needed. The term b thus represents a "mass source," which the pressure corrections (through their associated velocity corrections) must annihilate.

By now we have formulated all the equations needed for obtaining the velocity components and pressure. We are in a position to take an overall look at the entire solution algorithm.

6.7 THE SIMPLE ALGORITHM

The procedure that we are developing for the calculation of the flow field has been given the name SIMPLE, which stands for *S*emi-*I*mplicit *M*ethod for *P*ressure-*L*inked *E*quations. We shall discuss the significance of the name a little later. The procedure has been described in Patankar and Spalding (1972), Caretto, Gosman, Patankar, and Spalding (1972), and Patankar (1975).

6.7-1 Sequence of Operations

The important operations, in the order of their execution, are:

1. Guess the pressure field p^*.
2. Solve the momentum equations, such as Eqs. (6.8)–(6.10), to obtain u^*, v^*, w^*.
3. Solve the p' equation.
4. Calculate p from Eq. (6.11) by adding p' to p^*.
5. Calculate u, v, w from their starred values using the velocity-correction formulas (6.17)–(6.19).
6. Solve the discretization equation for other ϕ's (such as temperature, concentration, and turbulence quantities) if they influence the flow field through fluid properties, source terms, etc. (If a particular ϕ does not influence the flow field, it is better to calculate it after a converged solution for the flow field has been obtained.)
7. Treat the corrected pressure p as a new guessed pressure p^*, return to step 2, and repeat the whole procedure until a converged solution is obtained.

6.7-2 Discussion of the Pressure-Correction Equation

It will be recalled that in Section 6.5 we decided to drop the term $\Sigma\, a_{nb} u'_{nb}$ on our way to the velocity-correction formula (6.17). It is now time to explain the motivation for this and to affirm that no ultimate harm is entailed by this action.

1. If expressions such as $a_{nb}u'_{nb}$ were retained, they would have to be expressed in terms of the pressure corrections and the velocity corrections at the neighbors of u_{nb}. These neighbors would, in turn, bring their neighbors, and so on. Ultimately, the velocity-correction formula would involve the pressure correction at all grid points in the calculation domain, and the resulting pressure-correction equation would become unmanageable. We would, in effect, be going toward the direct solution of the whole set of momentum and continuity equations—a route that we decided not to follow. The omission of the $\Sigma\, a_{nb}u'_{nb}$ term enables us to cast the p' equation in the same form as the general ϕ equation, and to adopt a sequential, one-variable-at-a-time, solution procedure.

2. The words semi-implicit in the name SIMPLE have been used to acknowledge the omission of the term $\Sigma\, a_{nb}u'_{nb}$. This term represents an indirect or implicit influence of the pressure correction on velocity; pressure corrections at nearby locations can alter the neighboring velocities and thus cause a velocity correction at the point under consideration. We do not include this influence and thus work with a scheme that is only partially, and not totally, implicit.

3. The omission of any term would, of course, be unacceptable if it meant that the ultimate solution would not be the true solution of the discretized forms of the momentum and continuity equations. It so happens that the converged solution given by SIMPLE does not contain any error resulting from the omission of $\Sigma\, a_{nb}u'_{nb}$. In the converged solution, we acquire a pressure field such that the corresponding starred velocity field does satisfy the continuity equation. The details of the construction of the p' equation then become irrelevant to the correctness of the converged solution.

4. It is useful to focus attention on the operations during the "final" iteration, after which we are going to declare convergence. We have, as a result of all the previous iterations, come to possess a certain pressure field. Using this as p^*, we solve the momentum equations to get u^*, v^*, w^*. From this velocity field, we calculate the mass source b for the pressure-correction equation. Since this is going to be the final iteration, the value of b will come out to be practically zero for all the control volumes. Then, $p' = 0$ at all grid points will be an acceptable solution of Eq. (6.22), and the starred velocities and pressure will themselves be the correct velocities and pressure. Thus, the fact that the mass source b is zero everywhere is sufficient evidence that we have acquired the correct pressure field, and the actual solution of the p' equation is not needed during the final iteration. Obviously, the converged solution is then uninfluenced by any approximations made in deriving the p' equation—an equation that we really did not use in the final iteration.

5. The mass source b thus serves as a useful indicator of the convergence

of the fluid-flow solution. The iterations should be continued until the value of b everywhere becomes sufficiently small.

6. With this understanding, the pressure-correction equation can be seen to be merely an intermediate algorithm that leads us to the correct pressure field, but that has no direct effect on the final solution. As long as we get a converged solution, all formulations of the p' equation will give the same final solution.

7. The *rate* of convergence of the procedure will, however, depend on the particular formulation of the p' equation. If too many terms are omitted, divergence may result.

8. The pressure-correction equation derived in Section 6.6 is also prone to divergence unless some underrelaxation is used. Many different under-relaxation practices can be devised. A generally successful practice can be described as follows: We underrelax u^*, v^*, w^* (with respect to the previous iteration values of u, v, w) while solving the momentum equations [with a relaxation factor α, introduced in Eq. (4.55), set equal to about 0.5]; further, we add only a fraction of p' to p^*. In other words, instead of using Eq. (6.11), we employ

$$p = p^* + \alpha_p p' , \tag{6.24}$$

with α_p set equal to about 0.8. The task of Eq. (6.24) is to calculate p, which will be used as p^* in the next iteration; we can, in the interest of convergence, take any liberties in adjusting p^*. (The values of the relaxation factors that are mentioned here, namely $\alpha = 0.5$ and $\alpha_p = 0.8$, have been found to be satisfactory in a large number of fluid-flow computations. However, it is not implied that these values are the optimum ones or will even produce convergence for all problems. It should be recognized that matters such as the optimum relaxation-factor values are usually problem-dependent. Although experience from previous computations is helpful, new problems sometimes require different relaxation practices.)

9. It will be noticed that during each iteration the velocities are not left in their starred condition but are corrected using the velocity-correction formulas. The resulting velocity field exactly satisfies the discretized continuity equation, irrespective of the fact that the underlying pressure corrections are only approximate. Thus, the computations proceed to convergence via a series of *continuity-satisfying* velocity fields. This feature of SIMPLE has many advantages. A continuity-satisfying velocity field is likely to be more reasonable than the starred velocities. The use of underrelaxation with respect to these reasonable velocities helps in keeping the starred velocities also reasonable, and the mass sources small. Furthermore, the solution of the other ϕ equations in every iteration can be based on a flow field that satisfies a mass balance. To realize these advantages, one precaution is necessary: The velocity corrections should not be underrelaxed.

10. In the derivation of the p' equation, we considered the density ρ as known; the effect of pressure on density was not included. This can be regarded as a further approximation in the p' equation and justified in a similar manner. After all, this is the essence of any iterative method, which focuses attention on a few significant influences in the equation and regards many other quantities as tentatively known but to be recalculated for the next iteration. The density ρ is, in general, to be calculated from an appropriate equation of state. This may involve a dependence on temperature, concentration, and even pressure. As long as a converged solution can be obtained, our approximate p' equation is sufficient. For highly compressible (especially supersonic) flows, however, the dependence of density on pressure is so significant that there is a strong possibility of divergence. For such situations, it is desirable to derive a "compressible" form of the p' equation. This derivation has been set aside as an exercise (Problem 6.6).

11. It can be observed that the p' equation is very much like the discretization equation for heat conduction. In the velocity-correction formula (6.15), the velocity correction u'_e can be regarded as a heat flux caused by the temperature difference $p'_P - p'_E$.

12. The conductionlike nature of the p' equation implies that it does not exhibit one-way behavior in any space coordinate. It is well known that the influence of pressure is two-way or elliptic. The one-way behavior in boundary-layer flows is achieved by making an additional assumption about the pressure field; for example, the pressure variation normal to a wall is ignored in a wall boundary layer. Supersonic flows do exhibit one-way behavior in that the downstream pressure does not alter the upstream conditions. Computationally, we should use the compressible form of the p' equation (Problem 6.6) for supersonic flows. The coefficients in this form are similar to those in our convection-diffusion formulation, and then they do imply one-way behavior under appropriate Mach-number conditions.

It is interesting to note such close correspondence between theoretically established behavior and computational implications.

6.7-3 Boundary Conditions
for the Pressure-Correction Equation

The momentum equations are special cases of the general ϕ equation, and therefore our general boundary-condition treatment applies to them as well. However, since the p' equation is not one of the basic equations, some comments on the handling of its boundary conditions are appropriate.

Normally, there are two kinds of conditions at a boundary. Either the pressure at the boundary is given (and the velocity is unknown) or the velocity component normal to the boundary is specified.

Given pressure at the boundary. If the guessed pressure field p^* is arranged such that at a boundary $p^* = p_{given}$, then the value of p' at the

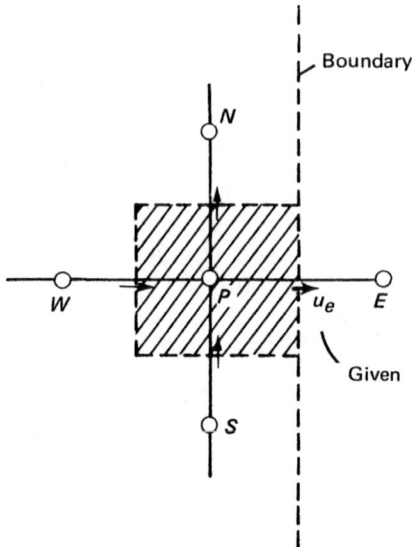

Figure 6.10 Boundary control volume for the continuity equation.

boundary will be zero. This is then akin to the given-temperature boundary condition in a heat-conduction problem.

Given normal velocity at the boundary. If the grid is designed such that the boundary coincides with a control-volume face, the situation will look like the one shown in Fig. 6.10. The velocity u_e is given. In the derivation of the p' equation for the control volume shown, the flow rate across the boundary face should *not* be expressed in terms of u_e^* and a corresponding correction, but in terms of u_e itself. Then, p'_E will not appear, or a_E will be zero in the p' equation. Thus, no information about p'_E will be needed.

6.7-4 The Relative Nature of Pressure

The foregoing description of the p' boundary conditions leads to a subtle but important issue. Let us consider a *constant-density* steady situation, in which the normal velocities are given at *all* boundary locations. Since no boundary pressure is specified and all the boundary coefficients such as a_E will be zero, the p' equation is left without any means of establishing the absolute value of p'. The coefficients of the p' equation are such that $a_P = \Sigma a_{nb}$ [see Eq. (6.23g)]; this means that p' and $p' + C$ (C is an arbitrary constant) would both satisfy the p' equation.

The situation, however, presents no real difficulty. For such a problem (in which the density is unaffected by pressure), the absolute value of pressure—and hence of pressure correction—is not relevant at all; only *differences* in pressure are meaningful, and these are not altered by an arbitrary constant added to the p' field. Pressure is then a *relative* variable, not an absolute one.

If the absolute value of p' is not unique, would the computations converge at all? Fortunately, an iterative method of solving the algebraic equations does converge to a solution, the absolute value of which is decided by the initial guess. A direct method, however, would encounter a singular matrix and refuse to give a solution. The remedy then is to arbitrarily assign the value of p' in one control volume and solve the p' equations for the *remaining* control volumes. The same technique can be used in an iterative method, but letting p' seek its own level gives faster convergence than insisting on a definite value at a certain point (see Problem 4.9).

Another way of looking at the indeterminate p' field is to note that the continuity equations for all the control volumes do not represent a linearly independent set. Since, in a properly specified problem, the given boundary velocities must satisfy *overall* mass conservation, the continuity equation for the last control volume does not convey any information that is not already contained in the continuity equations for all other control volumes. Thus, even if one of the control-volume equations is discarded (and the value of p' is prescribed there), the resulting corrected velocity field would satisfy continuity for *all* control volumes.

In many problems, the value of the absolute pressure is much larger than the local differences in pressure that are encountered. If the absolute values of pressure were used for p, round-off errors would arise in calculating differences like $p_P - p_E$. It is, therefore, best to set $p = 0$ as a reference value at a suitable grid point and to calculate all other values of p as pressures *relative* to the reference value. Similarly, before the p' equation is solved during each iteration, it is useful to start from $p' = 0$ as the guess for all points, so that the solution for p' does not acquire a large absolute value.

When the pressure at some boundary points is specified, or when the density depends on pressure, the indeterminancy of the pressure level does not arise.

6.8 A REVISED ALGORITHM: SIMPLER

The SIMPLE algorithm has been extensively used and has served well. For example, all the fluid-flow calculations to be presented in Chapter 9 were performed using this algorithm. However, in attempts to improve its rate of convergence, a revised version has been worked out. It is called SIMPLER, which stands for SIMPLE Revised (Patankar, 1979a).

6.8-1 Motivation

The approximation introduced in the derivation of the p' equation (the omission of the term $\Sigma a_{nb} u'_{nb}$) leads to rather exaggerated pressure corrections, and hence underrelaxation becomes essential. Since the influence of the neighbor-point velocity corrections is removed from the velocity-correction formula, the pressure correction has the entire burden of correcting the velocities, and this results into a rather severe pressure-correction field. In most cases, it is reasonable to suppose that the pressure-correction equation does a fairly good job of correcting the *velocities*, but a rather poor job of correcting the *pressure*.

To appreciate this argument, let us consider a very simple problem, one in which there is one-dimensional constant-density flow with the velocity given at the inlet boundary. It is easy to see that the velocity in this problem is governed only by continuity, and hence the continuity-satisfying velocity field obtained at the end of the first iteration will itself be the final answer. The predicted pressure, however, will be far from the final solution, owing to the approximate nature of the p' equation. It would take many iterations before a converged pressure field were established, although the correct velocity field is obtained very early in the process.

If we employ the pressure-correction equation only for the task of correcting the velocities and provide some other means of obtaining an improved pressure field, we construct a more efficient algorithm. This is the essence of SIMPLER.

6.8-2 The Pressure Equation

An equation for obtaining the pressure field can be derived as follows: The momentum equation (6.6) is first written as

$$u_e = \frac{\Sigma\, a_{nb} u_{nb} + b}{a_e} + d_e(p_P - p_E)\,, \qquad (6.25)$$

where d_e has been defined in Eq. (6.16). Now we define a pseudovelocity $\hat{u}_e$ by

$$\hat{u}_e = \frac{\Sigma\, a_{nb} u_{nb} + b}{a_e}\,. \qquad (6.26)$$

It can be noted that $\hat{u}_e$ is composed of the neighbor velocities u_{nb} and contains no pressure. Equation (6.25) now becomes

$$u_e = \hat{u}_e + d_e(p_P - p_E)\,. \qquad (6.27)$$

Similarly, we can write

$$v_n = \hat{v}_n + d_n(p_P - p_N)\,, \qquad (6.28)$$

$$w_t = \hat{w}_t + d_t(p_P - p_T)\,. \qquad (6.29)$$

It is easy to see the similarity between these equations and Eqs. (6.17)–(6.19). Here, $\hat{u}$, $\hat{v}$, $\hat{w}$ appear in place of u^*, v^*, w^*, and the pressure p itself takes the place of p'. It then follows that, if the derivation in Section 6.6 were worked out with the new velocity-pressure relations containing $\hat{u}$, $\hat{v}$, $\hat{w}$, an equation for pressure would result. This can be written as

$$a_P p_P = a_E p_E + a_W p_W + a_N p_N + a_S p_S + a_T p_T + a_B p_B + b, \quad (6.30)$$

where a_E, a_W, a_N, a_S, a_T, a_B, and a_P are given by Eqs. (6.23a)–(6.23g), and b is given by

$$b = \frac{(\rho_P^0 - \rho_P) \, \Delta x \, \Delta y \, \Delta z}{\Delta t} + [(\rho \hat{u})_w - (\rho \hat{u})_e] \, \Delta y \, \Delta z$$

$$+ [(\rho \hat{v})_s - (\rho \hat{v})_n] \, \Delta z \, \Delta x + [(\rho \hat{w})_b - (\rho \hat{w})_t] \, \Delta x \, \Delta y. \quad (6.31)$$

It should be noted that the expression for b is the only difference between the pressure equation (6.30) and the pressure-correction equation (6.22). Expression (6.31) for b uses the pseudovelocities $\hat{u}$, $\hat{v}$, $\hat{w}$, while b for the p' equation was calculated in terms of the starred velocities.

Although the pressure equation and the pressure-correction equation are almost identical, there is one major difference: No approximations have been introduced in the derivation of the pressure equation. Thus, if a correct velocity field were used to calculate the pseudovelocities, the pressure equation would at once give the correct pressure.

6.8-3 The SIMPLER Algorithm

The revised algorithm consists of solving the pressure equation to obtain the pressure field and solving the pressure-correction equation only to correct the velocities. The sequence of operations can be stated as:

1. Start with a guessed velocity field.
2. Calculate the coefficients for the momentum equations and hence calculate $\hat{u}$, $\hat{v}$, $\hat{w}$ from equations such as Eq. (6.26) by substituting the values of the neighbor velocities u_{nb}.
3. Calculate the coefficients for the pressure equation (6.30), and solve it to obtain the pressure field.
4. Treating this pressure field as p^*, solve the momentum equations to obtain u^*, v^*, w^*.
5. Calculate the mass source b [Eq. (6.23h)] and hence solve the p' equation.
6. Correct the velocity field by use of Eqs. (6.17)–(6.19), but *do not* correct the pressure.
7. Solve the discretization equations for other ϕ's if necessary.
8. Return to step 2 and repeat until convergence.

6.8-4 Discussion

1. It is easy to see that, for the one-dimensional problem discussed in Section 6.8-1, the SIMPLER algorithm would at once give a converged solution. In general, since the pressure-correction equation produces reasonable velocity fields, and the pressure equation works out the direct consequence (without approximation) of a given velocity field, convergence to the final solution should be much faster.

2. In SIMPLE, a guessed pressure field plays an important role. On the other hand, SIMPLER does not use guessed pressures, but extracts a pressure field from a given velocity field.

3. If the given velocity field happens to be the *correct* velocity field, then the pressure equation in SIMPLER will produce the correct pressure field, and there will be no need for any further iterations. If, on the other hand, the same correct velocity field and a guessed pressure field were used to start the SIMPLE procedure, the situation would actually deteriorate at first. The use of the guessed pressure would lead to starred velocities that would be different from the given correct velocities. Then, the approximations in the p' equation would produce incorrect velocity and pressure fields at the end of the first iteration. Convergence would take many iterations, despite the fact that we did have the correct velocity field at the beginning.

4. Because of the close similarity between the pressure equation and the pressure-correction equation, the discussion in Section 6.7-3 about boundary conditions for the p' equation is also relevant to the pressure equation. Furthermore, the relative nature of the pressure discussed in Section 6.7-4 could have been described by reference to the pressure equation.

5. Although SIMPLER has been found to give faster convergence than SIMPLE, it should be recognized that one iteration of SIMPLER involves more computational effort. First, the pressure equation must be solved in addition to all the equations solved in SIMPLE; and second, the calculation of $\hat{u}$, $\hat{v}$, $\hat{w}$ represents an effort for which there is no counterpart in SIMPLE. However, since SIMPLER requires fewer iterations for convergence, the additional effort per iteration is more than compensated by the overall saving of effort.

6.9 CLOSURE

In this chapter, we have completed the final step in constructing our numerical method. A number of miscellaneous, but important, topics still remain to be discussed. Although these could have been included in the first six chapters, they can be better appreciated at this stage, when the reader has a complete view of the procedure. The next chapter is devoted to these topics.

PROBLEMS

6.1 A two-dimensional flow with constant density and viscosity is governed by

$$\rho \frac{\partial u}{\partial t} + \rho u \frac{\partial u}{\partial x} + \rho v \frac{\partial u}{\partial y} = \mu \left(\frac{\partial^2 u}{\partial x^2} + \frac{\partial^2 u}{\partial y^2} \right) - \frac{\partial p}{\partial x} \, ,$$

$$\rho \frac{\partial v}{\partial t} + \rho u \frac{\partial v}{\partial x} + \rho v \frac{\partial v}{\partial y} = \mu \left(\frac{\partial^2 v}{\partial x^2} + \frac{\partial^2 v}{\partial y^2} \right) - \frac{\partial p}{\partial y} \, ,$$

and

$$\frac{\partial u}{\partial x} + \frac{\partial v}{\partial y} = 0 \, .$$

Eliminate p from the first two equations by differentiating the first with respect to y and the second with respect to x and subtracting one from the other. Express the resulting equation with ω as the dependent variable, where ω, the vorticity, is defined by $\omega \equiv \partial u / \partial y - \partial v / \partial x$. Show that the result is

$$\rho \frac{\partial \omega}{\partial t} + \rho u \frac{\partial \omega}{\partial x} + \rho v \frac{\partial \omega}{\partial y} = \mu \left(\frac{\partial^2 \omega}{\partial x^2} + \frac{\partial^2 \omega}{\partial y^2} \right) .$$

6.2 Define a stream function ψ as

$$\frac{\partial \psi}{\partial x} = -v \qquad \text{and} \qquad \frac{\partial \psi}{\partial y} = u \, .$$

Show that ψ identically satisfies the continuity equation given in Problem 6.1. Further, use the definition of ω in Problem 6.1 to show that

$$\frac{\partial^2 \psi}{\partial x^2} + \frac{\partial^2 \psi}{\partial y^2} = \omega \, .$$

6.3 In the steady, one-dimensional, constant-density situation shown in Fig. 6.11, the velocity u is calculated for locations A, B, and C, while the pressure p is calculated for locations 1, 2, and 3. The velocity-correction formula is

$$u = u^* + (p_i' - p_{i+1}')d \, ,$$

where the locations i and $i + 1$ lie on either side of the location for u. The value of d is 2 everywhere. The boundary conditions are $u_A = 10$ and $p_3' = 0$. If, at a given stage in the iteration process, the momentum equations give $u_B^* = 8$ and $u_C^* = 11$, calculate the values of p_1' and p_2'. Explain how you would obtain the values of p_1' and p_2' if the right-hand boundary condition were given as $u_C = 10$ instead of $p_3' = 0$.

6.4 A one-dimensional flow through a porous material is governed by $c|u|u + dp/dx = 0$, where c is a constant. The continuity equation is $d(uA)/dx = 0$, where A is the effective

Figure 6.11 Situation for Problems 6.3 and 6.4.

area for the flow. Use the SIMPLE procedure for the grid shown in Fig. 6.11 (where you may ignore point A) to calculate p_2, u_B, and u_C from the following data:

$$x_2 - x_1 = x_3 - x_2 = 2$$

$$c_B = 0.25 \qquad c_C = 0.2 \qquad A_B = 5 \qquad A_C = 4 \qquad p_1 = 200 \qquad p_3 = 38$$

As an initial guess, set $u_B = u_C = 15$ and $p_2 = 120$.

6.5 The one-dimensional flow in the nozzle shown in Fig. 6.12 can be described by

$$\frac{d}{dx}(\rho u A) = 0 \qquad \text{and} \qquad \frac{d}{dx}(\rho u A)u = -A\frac{dp}{dx}$$

where A is the cross-sectional area. The given conditions are

$$\rho = 1 \text{ everywhere} \qquad A_A = 3 \qquad A_B = 1 \qquad p_1 = 28 \qquad p_3 = 0.$$

Assume that the fluid upstream of point 1 has negligible momentum. Formulate the discretization equations for u and p', and hence obtain the values of u_A, u_B, and p_2. (Use the initial guesses $\rho u A = 5$, so that $u_A = \frac{5}{3}$ and $u_B = 5$, and $p_2 = 25$. Employ appropriate underrelaxation if necessary.)

6.6 Consider the steady, one-dimensional, *compressible* flow for which the continuity equation is $d(\rho u)/dx = 0$. With reference to Fig. 6.1, write the discretization form of this equation in terms of ρ_e, ρ_w, u_e, and u_w. Further, assume the density-correction formula $\rho = \rho^* + Kp'$, which can be derived from the appropriate equation of state. Assuming a piecewise-linear profile for p', derive the discretization equation for pressure correction. *Hint*: use the approximation

$$\rho u = (\rho^* + \rho')(u^* + u') \approx \rho^* u^* + \rho' u^* + \rho^* u'.$$

Note that the resulting coefficients have a convective part and a diffusive part, and that there is a possibility that the coefficients may become negative when the Mach number is large. Can you suggest an upwindlike scheme to prevent the coefficients from becoming negative?

6.7 A portion of a water-supply system is shown in Fig. 6.13. The flow rate Q in a pipe

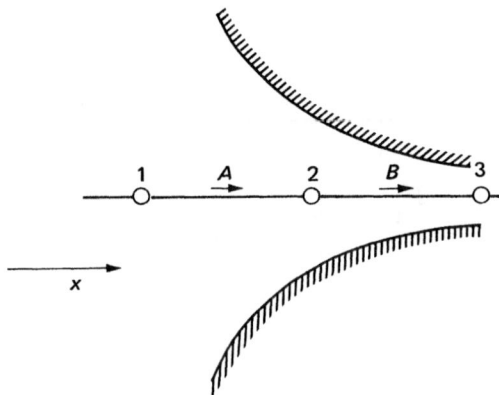

Figure 6.12 Grid points for Problem 6.5.

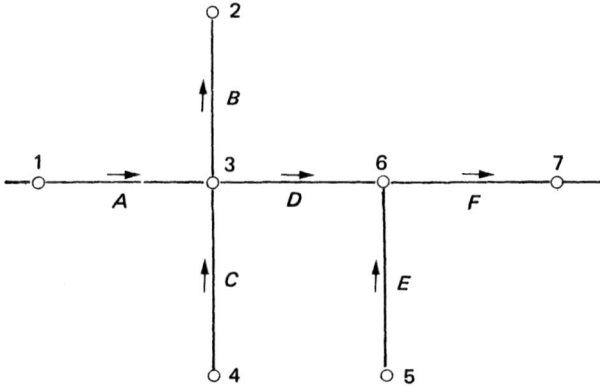

Figure 6.13 Water-supply system considered in Problem 6.7.

is given by $Q = C \, \Delta p$, where Δp is the pressure drop over the length of the pipe, and C is the hydraulic conductance. We have the following data:

$$p_1 = 275 \qquad p_2 = 270 \qquad p_4 = 0 \qquad p_5 = 40 \qquad Q_F = 20$$

$$C_A = 0.4 \qquad C_B = C_D = C_F = 0.2 \qquad C_C = C_E = 0.1 \, .$$

Find p_3, p_6, Q_A, Q_B, Q_C, Q_D, and Q_E by the following procedure: Guess p_3 and p_6. Obtain Q^* values based on the guessed pressures. Construct the pressure-correction equations and solve for p'_3 and p'_6. Correct the guessed pressures and the Q^* values. Do you need to iterate? Why?

SEVEN

FINISHING TOUCHES

7.1 THE ITERATIVE NATURE
OF THE PROCEDURE

The calculation procedure described in this book is aimed at solving coupled nonlinear equations by an iterative scheme. At this point, we shall take an overall look at the iterative process.

1. The iteration technique plays two different roles:

a. Our equations are, in general, nonlinear and interlinked. We cast them into nominally linear form and calculate the coefficients from the previous-iteration values of the variables.

b. The nominally linear algebraic equations for one dependent variable at a time are solved by an iterative method (such as the line-by-line method) rather than by a direct method.

2. The iterative solution of the algebraic equations need not be taken to complete convergence, because we are, at any intermediate stage, working with only tentative coefficients. After the discretization equations have been iterated to a certain extent, one must return to the recalculation of the coefficients. A sense of proportion is appropriate here. After having spent a certain amount of effort on calculating the coefficients, we must extract a fairly good solution of the algebraic equations, but refrain from doing an excessive amount of work with coefficient values that we know well to be only tentative. A direct solution method used for multidimensional problems

usually results in a disproportionately large amount of work spent in the equation-solving activity.

3. A similar consideration has been used in Chapter 6 in choosing a sequential, rather than simultaneous, procedure for calculating fluid flow. The momentum equations and the pressure-correction equation are solved sequentially. The alternative, which is commonly adopted in most finite-element methods for fluid flow, is to obtain a simultaneous solution of the linearized forms of the continuity equation and all the momentum equations. Such a simultaneous solution by a direct method requires large amounts of computer time and storage. Since the momentum equations are nonlinear, these large amounts of effort must be spent at every iteration. Further, the continuity and momentum equations may not be the only equations governing the situation. These equations are often coupled with the energy equation (through fluid properties and buoyancy forces), with the equations for turbulence parameters (through the turbulent viscosity), with the equations for chemical-species concentration, and so on. Obviously, it would not be practicable to attempt a simultaneous solution of all these equations; these additional equations would normally be solved in a sequential manner. Under these circumstances, the expenditure of large amounts of computing effort for the simultaneous solution of the continuity and momentum equations seems out of proportion.

4. In the numerical method presented in this book, there is no fundamental difference between solving a steady-state problem and performing one time step in an unsteady problem. In a steady problem, we start with guessed values for the variables ϕ and proceed to obtain the steady-state solution. For an unsteady situation, the problem is this: Given the values of ϕ at time t and a guess for ϕ at $t + \Delta t$, find the values of ϕ at $t + \Delta t$. As in the steady-state problem, we must perform a number of iterations at each time step for an unsteady problem. Further, many such time steps must be sequentially executed to cover the desired time period.

5. Thus, the solution of an unsteady problem seems to involve an effort that is equivalent to the task of solving a succession of steady-state problems. This is partially true, but there is one consolation. For reasonable values of Δt, the known ϕ values at time t can be used as a guess for the unknown ϕ values at time $t + \Delta t$. Since this is a relatively good guess (compared with a rather arbitrary guess, which one must make in a steady-state situation), only a few iterations are normally needed to obtain a converged solution for the time step. Sometimes, the number of iterations per time step can be as small as one. Thus, when a method for a nonlinear unsteady problem is claimed to be *noniterative*, it is, in fact, accepting the solution at the end of one iteration as a sufficiently converged solution for that time step. Such methods must employ rather small time steps, whereas the use of multiple iterations for a time step would allow larger values of Δt.

6. Such a one-iteration-per-time-step method is sometimes used to obtain

the steady-state solution at the end of many time steps. Such time steps are truly iterations, with the unsteady term in the equations providing a kind of underrelaxation.

7. A computer program that employs iteration within a time step should provide storage for the values of ϕ at time t and for the ϕ values at $t + \Delta t$. A steady-state program, on the other hand, requires storage for only one set of ϕ values, which are continually overwritten until convergence is attained.

8. The iterative technique greatly simplifies the construction of the numerical method and provides a way in which, at least in principle, one can handle any nonlinearity and interlinkage. Of course, the technique is of no value if a converged solution cannot be reached. It is useful at this stage to examine the prospects of convergence.

a. The four basic rules (introduced in Section 3.4) have enabled us to obtain such discretization equations as would, for fixed values of the coefficients, ensure convergence of the point-by-point or line-by-line solution procedure.

b. If the coefficients do not remain fixed but change rather slowly, it seems reasonable that we shall still obtain convergence. A proper linearization of the source term and an appropriate underrelaxation of the dependent variables would, in general, slow down the changes in the variables and hence in the coefficients.

c. In addition to the dependent variables, other quantities can be under-relaxed with advantage. For example, the density ρ is often the main link between the flow equations and the equations for temperature, concentration, etc. An underrelaxation of ρ via

$$\rho = \alpha \rho_{\text{new}} + (1 - \alpha)\rho_{\text{old}} \qquad (7.1)$$

would cause the velocity field to respond rather slowly to the changes in temperature and concentration. A diffusion coefficient Γ can be under-relaxed to restrain, for example, the influence of the turbulence quantities on the velocity field. The present value of Γ is then calculated from

$$\Gamma = \alpha \Gamma_{\text{new}} + (1 - \alpha)\Gamma_{\text{old}} \ . \qquad (7.2)$$

Here, as in Eq. (7.1), α stands for the relaxation factor. Underrelaxation requires α to be positive but less than 1. The interlinkage between different variables often comes through the source term (for example, the buoyancy force in a momentum equation depends on temperature). We may decide to underrelax the source term via

$$S_C = \alpha S_{C,\text{new}} + (1 - \alpha)S_{C,\text{old}} \ . \qquad (7.3)$$

Even the boundary conditions can be underrelaxed. For example, a hot wall or a rotating disc need not assume its final temperature or rotational speed right from the first iteration; the boundary value may be slowly adjusted, during the course of the iterations, to ultimately achieve the desired value. Thus,

$$\phi_B = \alpha\phi_{B,\text{given}} + (1 - \alpha)\phi_{B,\text{old}} \ . \tag{7.4}$$

Of course, the value of α appearing in Eqs. (7.1)-(7.4) need not be the same, nor is it necessary to use the same value for α for every grid point.

d. It must be remembered that there is no general guarantee that, for all nonlinearities and interlinkages, we will always get a converged solution. The underrelaxation procedures that are introduced here have been found to be helpful in many cases, but special underrelaxation practices may be needed for special problems. In the absence of an unconditional guarantee, we can nevertheless derive hope from the fact that, for a large number of rather complex problems, it has been possible to get converged solutions. A sample of such solutions will be presented in Chapter 9, but many other problems have also been solved and published.

9. As we have noted, an iterative process is said to have converged when further iterations will not produce any change in the values of the dependent variables. In practice, the iterative process is terminated when some arbitrary convergence criterion is satisfied. An appropriate convergence criterion depends on the nature of the problem and on the objectives of the computation. A common procedure is to examine the most significant quantities given by the solution (such as the maximum velocity, total shear force, a certain pressure drop, or overall heat flux) and to require that the iterations be continued only until the relative change in these quantities between two successive iterations is greater than a certain small number. Often the relative change in the grid-point values of all the dependent variables is used to formulate the convergence criterion. This type of criterion can sometimes be misleading. When heavy underrelaxation is used, the change in the dependent variables between successive iterations is intentionally slowed down; this may create an illusion of convergence although the computed solution may be far from being converged. A more meaningful method of monitoring convergence is to examine how perfectly the discretization equations are satisfied by the current values of the dependent variables. For each grid point, a residual R can be calculated from

$$R = \Sigma a_{\text{nb}}\phi_{\text{nb}} + b - a_P\phi_P \ . \tag{7.5}$$

Obviously, when the discretization equation is satisfied, R will be zero. A suitable convergence criterion is to require that the largest value of $|R|$ be less than a certain small number. Incidentally, as mentioned in Section 6.7.2, the quantity b in Eq. (6.22), which is the residual of the continuity equation, can be used as one of the indicators of the convergence of the iterative process.

7.2 SOURCE-TERM LINEARIZATION

In Section 4.2-5, the concept of the linearization of the source term was introduced. One of the basic rules (Rule 3) required that when the source term is linearized as

$$S = S_C + S_P \phi_P , \tag{7.6}$$

the quantity S_P must not be positive. Now, we return to the topic of source-term linearization to emphasize that often source terms are the cause of divergence of iterations and that proper linearization of the source term frequently holds the key to the attainment of a converged solution.

7.2-1 Discussion

1. It is important to watch for unintentional violations of the negative-S_P requirement. For example, in $r\theta z$ coordinates, the momentum equation for V_θ contains a source term $-\rho V_r V_\theta / r$. It is tempting to express this as $S_C = 0$ and $S_P = -\rho V_r / r$. However, if V_r happens to be negative, this gives a positive value of S_P. A proper formulation would be

$$S_C = \left[\!\!\left[\frac{-\rho V_r}{r}, 0 \right]\!\!\right] V_\theta , \tag{7.7a}$$

$$S_P = -\left[\!\!\left[\frac{\rho V_r}{r}, 0 \right]\!\!\right] . \tag{7.7b}$$

where $[\![\ ]\!]$ denotes the larger of the quantities listed within.

2. It is always possible to make S_P equal to zero, and to set $S_C = S$. However, this is often not desirable. The effect of a large negative S_P is much like that of underrelaxation and is conducive to convergence. As described in Section 4.2-5, probably the best linearization is one that makes the straight line $S = S_C + S_P \phi_P$ a tangent to the true $S \sim \phi$ curve. To use a smaller magnitude of S_P is to fail to adequately anticipate the decrease in S with an increase in ϕ. To use a larger magnitude of S_P is to be too cautious (which may at times be a good policy) and probably slow down the convergence.

3. Because the source terms are often large, it is always useful to consider the extreme case in which the source term alone dominates the discretization equation. For such a case, we may write the discretization equation as

$$S_C + S_P \phi_P \approx 0 \, , \qquad (7.8)$$

which leads to the solution

$$\tilde{\phi}_P = -\frac{S_C}{S_P} \, . \qquad (7.9)$$

Here, $\tilde{\phi}_P$ denotes the limiting value of ϕ_P in the source-dominated situation. In Fig. 7.1, these ideas are graphically represented. If the value S^* pertains to a current value ϕ_P^*, the solution of the discretization equation will be the value $\tilde{\phi}_P$, which corresponds to the point where the $S_C + S_P \phi_P$ line meets the abscissa. If S_P has a larger magnitude, $\tilde{\phi}_P$ will be closer to ϕ_P^*. A small magnitude of S_P would imply a larger change in ϕ_P from ϕ_P^* to $\tilde{\phi}_P$. The underrelaxation effect of S_P is thus obvious.

4. Sometimes, the source-dominated situation can be used to design the linearization such that ϕ_P remains within reasonable limits. Suppose that, for the current value ϕ_P^*, we desire that the next-iteration value of ϕ_P be close to a given value $\tilde{\phi}_P$. This can be arranged through the linearization

$$S_C = \frac{S^* \tilde{\phi}_P}{\tilde{\phi}_P - \phi_P^*} \, , \qquad (7.10a)$$

$$S_P = -\frac{S^*}{\tilde{\phi}_P - \phi_P^*} \, . \qquad (7.10b)$$

The desired value $\tilde{\phi}_P$ should be determined from physical considerations. For example, let ϕ stand for the mass fraction m_l of a chemical species. By

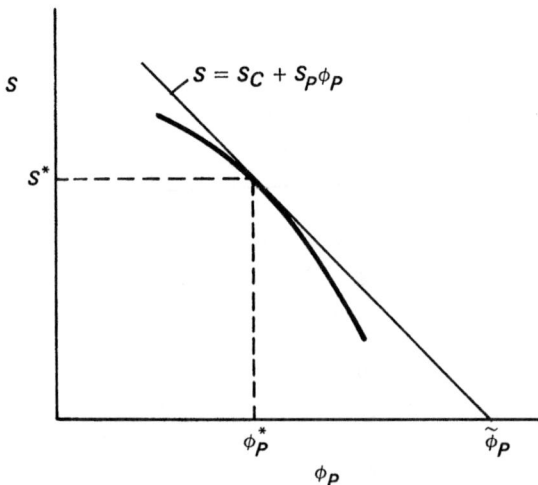

Figure 7.1 Solution in the source-dominated situation.

definition, m_l must lie between 0 and 1. At a current value m_l^*, if S^* is positive, m_l will increase and we may set $\tilde{m}_l$ as 1. For negative S^*, $\tilde{m}_l$ may be set equal to zero. We may wish to be even more conservative and require that in one iteration m_l could move only halfway toward the physical limit. Thus, $\tilde{m}_l$ would be set as $(m_l^* + 1)/2$ for positive S^*, and as $m_l^*/2$ for negative S^*. Because all these considerations are based on the source-dominated limit, the next-iteration value will not be exactly $\tilde{\phi}_P$, since the other terms in the equation also influence it. Further, we are not controlling the ultimate solution for ϕ_P, but simply its progress through the successive iterations. We seek to avoid rapid changes and physically unrealistic values from arising during the iterative process.

5. Normally, one is able to assign a known value of ϕ only at the boundary points. However, any desired value of ϕ can be arranged to be the solution at an *internal* grid point by setting S_C and S_P for that point as

$$S_C = 10^{30}\phi_{P,\text{desired}} \ , \tag{7.11a}$$

$$S_P = -10^{30} \ , \tag{7.11b}$$

where 10^{30} denotes a number large enough to make the other terms in the discretization equation negligible. The consequence is that

$$S_C + S_P\phi_P \approx 0 \ , \tag{7.12}$$

$$\phi_P = -\frac{S_C}{S_P} = \phi_{P,\text{desired}} \ . \tag{7.13}$$

This procedure can be used to represent internal obstacles or islands in the calculation domain by inserting "internal" boundary conditions.

7.2-2 Source Linearization
for Always-Positive Variables*

From the physical significance of certain dependent variables, we can conclude that their values always remain positive. Examples of such "always-positive" variables are mass fractions of chemical species, turbulence kinetic energy, turbulence length scale, and radiation fluxes in a flux model of radiation.

---

*For many readers, this seemingly minor topic may turn out to contain the most valuable information in this book. In practical computations, it is quite common to encounter erroneous results such as negative mass fractions and negative turbulence kinetic energy. These have such a devastating effect on the rest of the calculation and on the success of the iterations that they must be prevented at all costs. Fortunately, prevention is possible and easy.

Since such quantities usually have both positive and negative source terms (i.e., generation and destruction), the net source term can often become negative. If this is not properly handled, the always-positive variable may acquire an erroneous negative value.

The basic rule about positive coefficients (Rule 2 in Section 3.4) is crucial to ensuring physically realistic results. A further requirement for always-positive variables is that S_C must always be positive (and, of course, S_P always negative). Strict adherence to this requirement guarantees that no negative values of ϕ will arise.

There are many ways of ensuring that S_C is positive. A simple prescription is as follows: Suppose that

$$S = S_1 - S_2 \qquad S_1 > 0, \quad S_2 > 0 , \qquad (7.14)$$

where S_1 is the positive part of the source term, and $-S_2$ is the negative part. Since

$$S = S_1 - \frac{S_2}{\phi_P} \phi_P , \qquad (7.15)$$

we set

$$S_C = S_1 \qquad (7.16a)$$

and

$$S_P = -\frac{S_2}{\phi_P^*} , \qquad (7.16b)$$

where ϕ_P^* is the current value of ϕ_P.

7.3 IRREGULAR GEOMETRIES

We have developed our numerical method by using a grid in Cartesian coordinates. Since practical problems do not always fit neatly into such a coordinate system, it is necessary to discuss how the method can be applied to irregularly shaped domains.

7.3-1 Orthogonal Curvilinear Coordinates

Our use of Cartesian coordinates has been motivated mainly by convenience and ease of presentation. There is, however, no essential difficulty in working out the same numerical method in cylindrical or spherical coordinates or even in general orthogonal curvilinear coordinates. This was briefly illustrated in Section 4.6-2 for the $r\theta$ coordinates. More generally, one can employ an

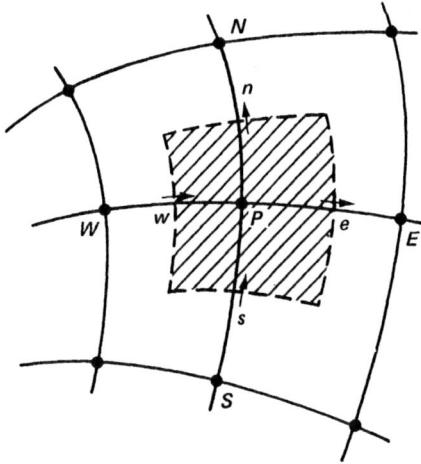

Figure 7.2 Control volume in an orthogonal curvilinear grid.

orthogonal curvilinear grid as shown in Fig. 7.2. In such a grid, the calculation of various lengths, areas, and volumes is not as straightforward as in a Cartesian grid, but otherwise all the practices developed for Cartesian grids are directly applicable.

The *orthogonal* property of the grid is, however, essential for the application of the method. Since we calculate a diffusion flux across a control-volume face in terms of the ϕ values at two grid points, it is crucial that the face is normal to the line joining the two grid points.

For an arbitrarily shaped domain, the construction of an orthogonal curvilinear coordinate system is itself a substantial task. Some procedures for doing this are now available [for example, Potter and Tuttle (1973)]. If the grid can be conveniently and economically constructed, then the use of orthogonal curvilinear coordinates is a viable method for handling irregular geometries.

7.3-2 Regular Grid with Blocked-off Regions

Sometimes a computer program written for a regular grid (such as the Cartesian grid) can be improvised to handle an irregularly shaped calculation domain. This is done by rendering inactive, or "blocking-off," some of the control volumes of the regular grid so that the remaining active control volumes form the desired irregular domain. Some examples are shown in Fig. 7.3, where the shaded areas denote the inactive control volumes. It is obvious that the irregular boundary must be approximated by a series of rectangular steps, but often surprisingly good answers can be obtained from a rather crude representation of the boundary.

The blocking-off operation consists of establishing known values of the relevant ϕ's in the inactive control volumes. If the inactive region represents a

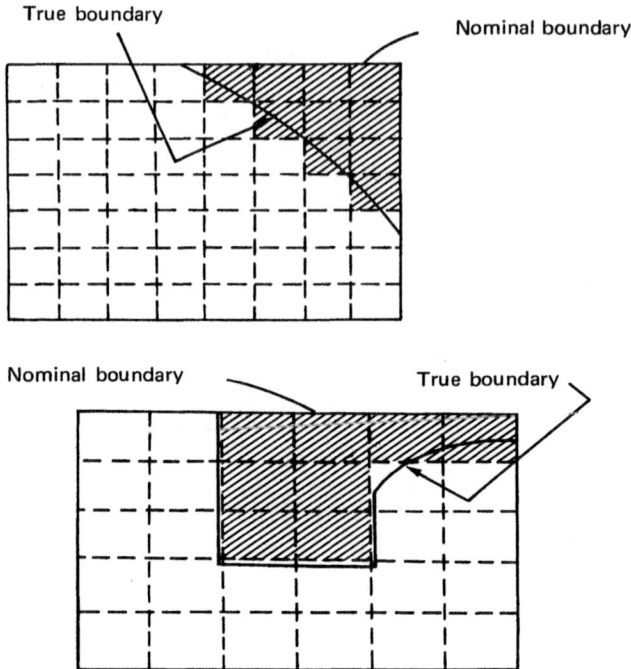

Figure 7.3 Blocked-off regions in a regular grid.

stationary solid boundary, the velocity components in that region must be set equal to zero. If the region is to be regarded as an isothermal boundary, the known temperature must be established in the inactive control volumes.

There are two ways in which the desired values can be set in the inactive control volumes. One method is the use of large source terms, as described in Section 7.2-1. An alternative is available through our use of harmonic-mean Γ's for the control-volume faces (Patankar, 1978), which was explained in Section 4.2-3. Since large discontinuities of Γ can be correctly handled, a very large value of Γ in the inactive zone would ensure that the value prescribed at the (nominal) boundary of the zone prevails over the entire inactive zone. Yet, the solution in the active zone will be unaffected by these large values of Γ. In particular, the velocities in the inactive zone can be set to zero by the use of a very large viscosity for the zone and a zero value of velocity at the nominal boundary.

It should be noted that, by these means, only rather simple boundary conditions can be handled at an irregular boundary. More complex boundary conditions would require modification of the source terms for the active control volumes adjacent to the true boundary. Also, the blocking-off method is somewhat wasteful of computer time and storage, since trivial computations must be performed for the inactive zone, and the results have to be stored.

Notwithstanding these considerations, the convenience of using a regular-grid computer program for any arbitrary geometry offers a significant advantage.

A further spin-off of the harmonic-mean Γ's is the ability to handle conjugate heat transfer problems, which will be discussed next.

7.3-3 Conjugate Heat Transfer

Let us consider the situation shown in Fig. 7.4. The fluid flows through a duct with an internal fin. The duct wall and the fin have finite thickness and moderate conductivity. The thermal boundary condition is known at the *outer* surface of the wall as, for instance, a prescribed temperature for that surface. The situation presents a conjugate heat transfer problem in that conduction in the solid and convection in the fluid must both be considered, with a proper matching at the fluid-solid interface. The calculation of separate solutions for the solid and fluid regions would require an involved iterative procedure for matching the interface condition. The harmonic-mean practice for Γ offers a much easier alternative that has been described in Patankar (1978).

In this procedure, the problem is solved by using a calculation domain that includes both the fluid and solid regions, with the *outer* surface of the wall coinciding with the boundary of the domain. Thus, the boundary conditions for both the velocity and temperature fields can easily be supplied at the outer surface of the wall. The calculation procedure rests on our ability to handle a large step change in the value of Γ. When the velocity equations are solved, Γ for the grid points that fall in the fluid region is made equal to the viscosity of the fluid, while for the grid points lying in the solid region Γ is set equal to a very large number. This would ensure that the zero velocity specified at the outer surface of the wall would prevail throughout the solid region, and thus the fluid region would experience the correct boundary condition.

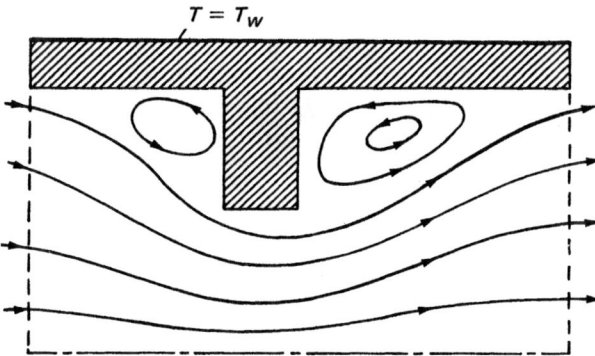

Figure 7.4 Conjugate heat transfer problem.

For solving the temperature equation, we specify the Γ field by employing the true conductivities of the solid and of the fluid in their respective regions. The problem is solved as a convection-conduction problem throughout the entire calculation domain; but, since the velocities in the solid are zero, the Peclet number there would also be zero, and, in effect, a pure-conduction calculation would be performed in the solid region. The resulting solution would thus give us temperature distributions in the solid and in the fluid, and they would have been automatically matched at the solid-fluid interface. As far as our calculation is concerned, this interface is simply an interior location, which is treated like any other interface between two control volumes.

7.4 SUGGESTIONS FOR COMPUTER-PROGRAM PREPARATION AND TESTING

To perform practical computations, the numerical method must be embodied in a computer program. It takes an organized and dedicated effort to produce an efficient and error-free program. After a computer program has been developed and tested, it becomes a valuable tool for the analyst. It opens up a whole new world of possibilities for solving complex practical problems with relative ease. The following suggestions are offered for the benefit of the readers who wish to develop computer programs for their needs.

1. The first step in the design of a computer program is to decide on the scope and limitations of the program. Will it handle two or three dimensions, Cartesian or cylindrical coordinates, uniform or nonuniform grids, constant or variable density, steady or unsteady problems? Too much generality makes the program voluminous and inconvenient to apply to simple problems. Too little generality restricts its use to a very few physical situations. Initially, it is probably best to develop a rather restricted version of the program with, however, a flexible framework so that the scope of the program can be easily enlarged.
2. It is useful to distinguish between general operations (such as the calculation of the coefficients and the solution of the discretization equations) and problem-dependent operations (such as the specification of Γ, S_C, S_P, and the boundary conditions for the relevant variables). The general operations should be programmed first and then tested with different problem specifications.
3. When a computer program is developed, it must be thoroughly tested. A program that contains errors is like a faulty instrument; it is unrealiable and misleading. It is possible to construct error-free computer programs, in which the analyst-programmer can take pride.
4. It is helpful to test separate parts of the program before the entire

assembly is put to work. For example, the subroutine for solving the discretization equations can be independently tested by supplying arbitrary values for the coefficients.

5. Most of the initial testing can be based on only coarse grids. This saves computer time, and, since the resulting fields of ϕ contain only a few numbers, it is easy to examine and interpret them. At times, some of the surprising results can be checked by manual calculation. Even the coarse-grid solutions are expected to be physically realistic, since this criterion has been the guiding principle in this book.

6. The control-volume approach ensures that the solution satisfies overall conservation over the calculation domain. Such overall balances provide a useful test of the computer program. In verifying overall balances, we must use the same profile assumptions as were used in constructing the discretization equations. Then, for a well-converged solution, overall conservation must be *perfectly* satisfied for any number of grid points. Alternatively, the overall balance may be taken as an indication of the satisfactoriness of the convergence of iterations.

7. To confirm the internal consistency of the computer program, a number of tests can be undertaken. One of them is to check that the converged solution is independent of the initial guesses and the relaxation factors.

8. The orientation of the coordinate system relative to the physical problem is, of course, arbitrary. The correctness of the computer program can be checked by solving the same problem by interchanging, for example, the x and y directions.

9. When the boundary conditions imply that the solution will be symmetrical about a line (or a plane), it is sufficient to perform the computation for only one-half of the domain lying on one side of the symmetry line. For example, the flow in a parallel-plate channel can be computed by using a *calculation* domain that extends from one plate to the center line between the plates. While testing the computer program, however, we can choose the whole domain (from one plate to the other) as the calculation domain and check whether the computed solution does exhibit the expected symmetry,[*] and whether the solution in each half is identical to the one obtained by using half the region as the calculation domain.

10. Suppose that the solution for a given problem is determined by the values of certain dimensionless parameters. For example, the Reynolds number $\mathrm{Re} = \rho U D / \mu$ may be the governing parameter. The solution for a

---

[*]There are some situations in which even with symmetrical boundary conditions the solution may not be symmetrical. For example, jets in ducts (which are used in fluidic devices) or sudden enlargements in duct flows often result in unsymmetrical flow patterns. Obviously, such special situations are not to be used in testing the program for symmetry.

specific value of Re can be obtained by setting, in the computer program,

$$\rho = 1 \qquad D = 1 \qquad \mu = 1 \qquad U = \text{Re} ,$$

or
$$U = 1 \qquad D = 1 \qquad \mu = 1 \qquad \rho = \text{Re} ,$$

or
$$\rho = 10 \qquad U = 5 \qquad \mu = 1 \qquad D = \frac{\text{Re}}{50} ,$$

or any other combination. The *dimensionless* outcome of the computed solution must be identical for all these combinations. This criterion can be used to verify the correctness of the computer program.

11. The principle of superposition, which is valid for linear heat conduction problems, can be used to test the consistency of the computer program. According to the superposition principle, the solutions for two rather simple problems can be added to construct the solution for a more complex problem. The computer program can be used directly to obtain the solutions for all three problems, and then it can be verified that the solution for the complex problem is indeed the sum of the solutions for the two other problems.

12. Limiting behavior under appropriate conditions provides a useful test of the computer program. A three-dimensional computer program can be employed to solve a two-dimensional problem to confirm that the computed solution is indeed two-dimensional. Computations for a duct flow should exhibit the expected fully developed behavior in the far-downstream region. A program for viscous flow should produce the inviscid solutions when the viscosity is set equal to zero.

13. The tests described so far have been aimed at checking the qualitative behavior of the computed results. Quantitative checks are also necessary, not only to confirm the correctness of the program but also to indicate the accuracy obtainable (with a certain grid fineness. Comparison with available exact solutions provides a useful way of testing the accuracy of the numerical solution. It should be verified that as the grid is refined the error in the computed solution diminishes. Since most standard exact solutions either deal with rather simple problems or require the calculation of infinite series involving special functions and eigenvalues, a method for constructing exact solutions is desirable. A convenient method is to *propose* a solution for ϕ, to provide the distributions of Γ, ρ, and u, and then to obtain an expression for S in Eq. (2.13) by substituting the other quantities into the equation. With this expression for S as the *given* source term (and with the given variations of Γ, ρ, and u), the proposed solution for ϕ can be regarded as the exact solution. Indeed, any domain over which ϕ is defined can be chosen as the calculation domain, and the

values of ϕ obtained from the exact solution at the boundaries of this domain can be used as the required boundary conditions.

14. Finally, published numerical solutions can be used to verify the correctness of a new computer program. For this purpose, the results of some of the illustrative applications presented in Chapter 9 will be useful.

EIGHT

SPECIAL TOPICS

In this book so far, a general method has been developed for the calculation of fluid flow, heat transfer, and related phenomena. Although one-dimensional and two-dimensional situations were used for ease of derivation and visualization, the ultimate treatment has dealt with the unsteady three-dimensional situation. Also, although the concept of a one-way space coordinate has been introduced, all the derivations have been based on two-way (i.e., elliptic) behavior for all the space coordinates.

The idea of a one-way space coordinate is, however, a very useful one, and special procedures that take advantage of one-way behavior have great practical utility. A few such procedures will be outlined in this chapter. Also, a finite-element method that uses many of the principles developed in this book will be briefly introduced. This will serve to emphasize the basic similarity between the finite-difference and finite-element approaches, which are often presented as entirely different methods.

This chapter is not intended as an exhaustive treatment of the topics chosen. The purpose of the chapter is to draw the attention of the reader to these special topics, which are closely related to the main theme of the book. With the background of this book and the cited references, the reader should be able to work out the required algebraic details.

8.1 TWO–DIMENSIONAL PARABOLIC FLOW

When a steady two-dimensional flow has one one-way space coordinate, it is called a *two-dimensional parabolic flow*. Such a flow has a predominant velocity

155

in the one-way coordinate, and hence the convection always dominates the diffusion in that coordinate. It is this feature that imparts the one-way character to the streamwise direction. Obviously, no reverse flow in that direction would be acceptable. A further requirement arises from the influence of pressure. It was indicated in Section 6.7-2 that pressure normally exerts two-way (or elliptic) influences. For the streamwise coordinate to be treated as one-way, the pressure variations in the cross-stream direction must be regarded as negligible.

Examples of two-dimensional parabolic flows are plane or axisymmetric cases of boundary layers on walls, duct flows, jets, wakes, and mixing layers. The solution for such situations is obtained by starting with a known distribution of ϕ at an upstream station and marching in the streamwise direction. For every forward step, the distribution of ϕ in the cross-stream coordinate is calculated at one streamwise station. Thus, computationally only a one-dimensional problem needs to be handled, for which the TDMA can be used to solve the discretization equations.

The solution of the momentum and continuity equations presents no special problem. The streamwise pressure gradient is assumed to be known. With this pressure gradient, the streamwise momentum equation is solved to yield the streamwise velocity. The cross-stream velocity is then calculated from the continuity equation. The pressure gradient for external flows comes from the pressure field in the external irrotational stream outside the boundary layer. For confined flows, overall mass conservation across the duct cross section is used to adjust the streamwise pressure gradient. No counterpart of SIMPLE or SIMPLER is needed for two-dimensional parabolic flows.

Complete details and computer programs for two-dimensional parabolic situations are available in Patankar and Spalding (1970) and Spalding (1977). The calculation method described therein uses a dimensionless stream function as the cross-stream coordinate, which provides a convenient way of expanding and contracting the width of the calculation domain in conformity with changes in the thickness of the boundary layer.

8.2 THREE–DIMENSIONAL PARABOLIC FLOW

If in a steady three-dimensional flow there exists one one-way coordinate, the flow can be characterized as a three-dimensional parabolic flow. Again, the conditions under which a space coordinate becomes one-way are the existence of a predominant unidirectional velocity in that coordinate; hence, negligible diffusion and absence of reverse flow in that direction; and negligible pressure variations in the cross-stream plane.

Examples of three-dimensional parabolic situations are similar to their two-dimensional counterparts. The boundary layer over a skewed airfoil, the flow in a duct of rectangular cross section, and a jet issuing from a noncircular orifice are all three-dimensional parabolic flows.

Although the apparent difference between the two- and three-dimensional parabolic situations is slight, the solution procedure needed for three-dimensional parabolic problems is far more complex than that for two-dimensional parabolic flows. (The SIMPLE procedure, it is worth noting, was first formulated in connection with three-dimensional parabolic flows in Patankar and Spalding, 1972a.) The reason is that, after the streamwise velocity has been calculated from the streamwise momentum equation, the *two* cross-stream velocities cannot be obtained from the continuity equation alone. To determine how the flow distributes itself in the two cross-stream directions, both cross-stream momentum equations must be solved. The two-dimensional parabolic procedure, on the other hand, does not employ the cross-stream momentum equation.

Because of the direct reference to cross-stream momentum equations, an assumption about pressure, which goes unnoticed in the procedure for two-dimensional parabolic flows, comes to the forefront in the three-dimensional parabolic procedure. This assumption is that the streamwise velocity is influenced by a cross-sectional mean pressure $\bar{p}$, while the cross-stream velocities are "driven" by a pressure variation p over the cross section. This pressure "decoupling" is essential to the use of a parabolic procedure.*

For external flows, the streamwise variation of $\bar{p}$ is obtained from the surrounding irrotational stream. In confined flows, the $\bar{p}$ variation is adjusted to satisfy overall mass conservation over the duct cross section. In a given forward step, once the streamwise velocity has been obtained with the appropriate streamwise gradient of $\bar{p}$, the problem of calculating the two cross-stream velocities and the cross-sectional pressure distribution is almost identical to a two-dimensional elliptic problem, which can be solved by the use of SIMPLE or SIMPLER. The details can be found in Patankar and Spalding (1972a), which can be easily interpreted with the background provided by this book.

8.3 PARTIALLY PARABOLIC FLOWS

In some practical situations there exists a predominant flow direction, and yet the cross-stream pressure variation is *not* negligible. Thus, the pressure decoupling employed in the parabolic procedures is not appropriate for such flows. In all other respects, the solution can be obtained by marching from the upstream end of the domain to the downstream end, but the downstream effects *are* transmitted upstream via pressure. Such situations are called

---

*The cross-sectional pressure p could be regarded as a perturbation over the mean pressure $\bar{p}$. For the flow to be treated as parabolic, the pressure perturbation over a cross section should be small so that, in the streamwise momentum equation, no significant error is introduced by the use of the mean pressure $\bar{p}$ instead of the actual local pressure.

partially parabolic. Highly curved ducts, a jet in a cross stream, ducts with a rapid change of cross section, and rotating passages are examples of partially parabolic situations. The basic concept of this class of flows was presented by Pratap and Spalding (1975, 1976), and has been applied to a film-cooling situation by Bergeles, Gosman, and Launder (1976, 1978).

In the partially parabolic calculation procedure, the pressure field is stored for the entire calculation domain, while all other variables are stored for only one or two marching stations. For a given pressure field, the marching procedure is employed just as in the fully parabolic situation, while an improved pressure field is obtained from a pressure-correction equation or a pressure equation. Many repetitions of the marching procedure are needed before a converged solution is obtained.

Compared with the fully elliptic procedure, the fully parabolic procedure offers savings in both computer time and computer storage. The *partially* parabolic procedure saves storage, but the savings in computer time may not be appreciable.

8.4 THE FINITE-ELEMENT METHOD

8.4-1 Motivation

The discretization method described in this book has, because of its use of regular grids, the appearance of a finite-difference method. In stress analysis, the finite-element method is much more commonly used than the finite-difference method; and, even in heat transfer and fluid flow, applications of the finite-element method have started appearing in increasing numbers.

The finite-element method subdivides the calculation domain into elements, such as the triangular elements shown in Fig. 8.1. The discretization equations are usually derived by the use of a variational principle when one exists or by the Galerkin method, which is a special case of the method of weighted residuals. In the derivation, a "shape function" or profile assumption is used to describe how the dependent variable ϕ varies over an element.

As explained in Section 3.2, the control-volume formulation is another special case of the method of weighted residuals. We also have used shape functions to describe the variation of ϕ between two grid points. It so happens that these shape functions have been locally one-dimensional; it is because of this feature that the grid lines are required to form an orthogonal net.

From this viewpoint, the finite-element method should not be considered as a basically different method. Its extra power lies only in its ability to use an irregular grid. Although we have discussed in Section 7.3 some ways of adapting our discretization method to irregular geometries, there is no doubt that the triangular grids shown in Fig. 8.1 provide more flexibility in fitting

SPECIAL TOPICS   **159**

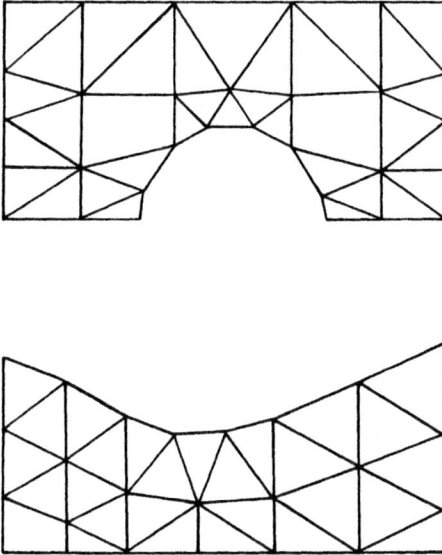

Figure 8.1 Examples of domain discretization by triangular elements.

irregular domains and in providing local grid refinement. The development of
a satisfactory finite-element method for heat transfer and fluid flow seems
highly desirable.

8.4-2 Difficulties

Although this potential of the finite-element method has been recognized for
quite some time, certain difficulties have, until recently, blocked progress:

1. The foremost difficulty concerns the upwind nature of convection. A
   straightforward application of the standard finite-element method would
   give a formulation that is very similar to the central-difference scheme; and
   we know too well that such a formulation can lead to physically unrealistic
   results. Something like the upwind or the exponential scheme is needed,
   but it is not clear how to adapt such a formulation to irregular grids.
2. The use of staggered grids was possible because the grid lines were laid out
   along coordinate directions, and the velocity components in these directions could be appropriately displaced. The need for something like a
   staggered grid is present in the triangular grid too; if all the variables were
   to be calculated for the same grid points, difficulties similar to those
   discussed in Section 6.2 would certainly arise.
3. Most of the published applications of the finite-element method to fluid
   flow employ a direct simultaneous solution of the continuity equation and
   all the momentum equations to yield the velocity components and
   pressure. Since direct solutions are expensive, it is desirable to formulate a

SIMPLElike sequential—rather than simultaneous—solution of the momentum and continuity equations.

4. For most fluid-flow–heat-transfer practitioners, the finite-element method still has a veil of mystery about it. The variational formulation, or even the Galerkin approach, does not have an easy physical interpretation. In conformity with the philosophy adopted in this book, it is desirable to produce a version of the finite-element method in which the physical meaning of the discretization equations can be readily understood.

8.4-3 A Control-Volume-based Finite-Element Method

The recent work of Baliga and Patankar (1979a, 1979b) has been successful in removing the aforementioned difficulties, and a finite-element method that is closely related to the discretization method described in this book has been formulated. The actual formulation was worked out for a two-dimensional situation, but care was taken to ensure that the extension to three dimensions can be made without the need for any further novelties. A brief description of the salient features of the method now follows.

1. For the triangular grid the dependent variables are calculated for grid points that lie at the vertices of the triangles. The discretization equations are formed by the control-volume method; i.e., the differential equation is intergrated over the typical control volume shown in Fig. 8.2. The control volumes are constructed by joining the centroid of each triangle to the midpoints of the sides of that triangle. This construction of the control volume was earlier proposed by Winslow (1967). It can be seen from Fig. 8.2 that the triangular elements adjacent to the grid point P accommodate portions of the control volume and the corresponding control-volume faces. The discretization equation is formed by adding the contributions of these elements to the integral conservation for the control volume.

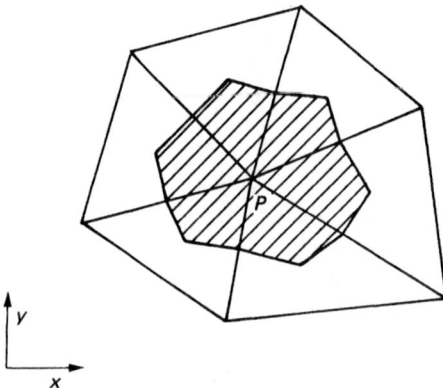

Figure 8.2 Control volume for the triangular grid.

2. A shape function describing the variation of ϕ over an element is needed to calculate the flux across the control-volume faces that fall within the element. The standard shape function for the triangular element is

$$\phi = a + bx + cy , \qquad (8.1)$$

where the constants a, b, c are expressed in terms of the three grid-point values of ϕ. For convection-diffusion problems, this shape function would give results much like the central-difference scheme in finite-difference methods. Since these results do become physically unrealistic when the Peclet number is large, the shape function given by Eq. (8.1) is unacceptable. The alternative proposed by Baliga and Patankar (1979a) is the shape function

$$\phi = A + B \exp \frac{\rho U X}{\Gamma} + CY , \qquad (8.2)$$

where U is the resultant velocity in the element, X is the coordinate pointing in the direction of the resultant velocity, and Y in the direction normal to it. The constants A, B, C are found in terms of the three values of ϕ at the vertices of the triangle.

On the basis of the discussion of convection and diffusion in Chapter 5, the rationale for the use of the exponential function in Eq. (8.2) should be quite obvious. For low Peclet numbers, Eq. (8.2) reduces to Eq. (8.1), which is the appropriate shape function for conduction problems. It is through the shape function (8.2) that the spirit of the exponential scheme has been introduced into the finite-element method.

In fact, the exponential shape function has achieved something more. Whereas the formulation in Chapter 5 uses locally one-dimensional representation, Eq. (8.2) works with the resultant-velocity direction. Consequently, the finite-element method based on Eq. (8.2) produces much less false diffusion than does the formulation in Chapter 5.

3. The issue of the staggered grid is handled by calculating the pressure

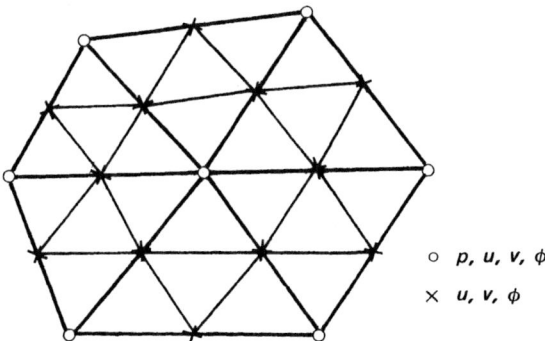

o p, u, v, ϕ

× u, v, ϕ

Figure 8.3 Macrotriangles and subtriangles.

on a grid that is different from the grid used for all the other variables. The pressure is calculated at the vertices of "macrotriangles," which are shown in Fig. 8.3 by small circles. Each macrotriangle is divided into four subtriangles. The subtriangles form the grid for the velocity components and all other variables except pressure.

4. A sequential solution algorithm in the spirit of SIMPLER is formulated. The pressure equation and the pressure-correction equation are derived from the continuity equation written for a control volume defined by the macrotriangle.

The control-volume-based finite-element method outlined here has undergone relatively little testing, and certainly numerous refinements can be made. However, the method represents a logical and effective extension of our discretization method to triangular grids.

ILLUSTRATIVE APPLICATIONS

In this last chapter, we shall look at a few applications of the numerical method described in this book. The method has been extensively tested and applied to a variety of practical situations. A review paper (Patankar, 1975) written in the early days of the SIMPLE procedure contains a number of examples that were available at that time. Since then, many more applications have appeared in the literature. A partial list of the published applications of the method now follows.

Two-dimensional elliptic situations involving fluid flow and heat transfer have been computed by Lilly (1976), Abdel-Wahed, Patankar, and Sparrow (1976), Moon and Rudinger (1977), Majumdar and Spalding (1977), Patankar, Liu, and Sparrow (1977), Durst and Rastogi (1977), Sparrow, Patankar, and Ramadhyani (1977), McGuirk and Rodi (1978), Patankar, Ramadhyani, and Sparrow (1978), Ganesan, Spalding, and Murthy (1978), Patankar, Sparrow, and Ivanović (1978), Sparrow, Patankar, and Shahrestani (1978), Sparrow, Baliga, and Patankar (1978), and Patankar, Ivanović, and Sparrow (1979).

Issa and Lockwood (1977) have modified the basic calculation method to handle both subsonic and supersonic regions in a single domain. Turbulent reacting flow in two-dimensional furnaces has been computed by Khalil, Spalding, and Whitelaw (1975). Patankar and Spalding (1972b, 1974b) have used the three-dimensional elliptic procedure for situations involving turbulence, combustion, and radiation. Other three-dimensional elliptic problems have been solved by Caretto, Gosman, Patankar, and Spalding (1972), Patankar and Spalding (1974a, 1978), and Patankar, Basu, and Alpay (1977).

The method for three-dimensional parabolic flows has been applied to complex practical problems by Patankar, Rastogi, and Whitelaw (1973), Patankar, Pratap, and Spalding (1974, 1975), Patankar, Rafiinejad, and Spalding (1975), McGuirk and Rodi (1977), Majumdar, Pratap, and Spalding (1977), Rostogi and Rodi (1978), and DeJoode and Patankar (1978).

A complete discussion of all these applications will not be attempted here. The aim of this chapter is to give the reader a feel for some applications and then leave the rest to the imagination. Since only a few applications would serve this purpose, it was convenient to choose them from the problems solved by the author and his co-workers.

It is interesting to note that all the applications presented here have been worked out by the use of only three general-purpose computer programs. The three computer programs differ only in their dimensionality and parabolic-elliptic nature. The programs are respectively designed for (1) two-dimensional elliptic situations, (2) three-dimensional parabolic situations, and (3) three-dimensional elliptic situations. It is possible to arrange each program to handle either the Cartesian or cylindrical coordinate system. Of course, the adaptation of any of the programs to a particular problem requires the incorporation of appropriate mathematical models for the relevant physical processes (such as turbulence or chemical reaction) and the introduction of the problem specifications (such as geometry, fluid properties, and boundary conditions). Although this adaptation often represents a significant effort, the use of general-purpose computer programs still provides a great convenience.

Among the eight examples presented in this chapter, those in Sections 9.4–9.6 involve turbulent flow. The standard k-ϵ model of turbulence (Launder and Spalding, 1974) is used in Sections 9.5 and 9.6, while a special version of the mixing-length model is employed in Section 9.4. The steam-generator problem in Section 9.8 employs the concept of distributed resistances for flow over a tube bundle. The remaining sections deal with laminar-flow situations.

From the computational point of view, a two-dimensional elliptic problem is involved in the situations treated in Sections 9.2–9.4 and 9.7; the problems in Sections 9.1 and 9.6 employ the three-dimensional parabolic procedure; and Sections 9.5 and 9.8 illustrate the application of the three-dimensional elliptic procedure. All the situations are steady-state except the one in Section 9.3, where a moving-boundary unsteady problem is handled.

9.1 DEVELOPING FLOW IN A CURVED PIPE

The axisymmetric flow in a straight circular pipe is two-dimensional in character. The flow in a curved pipe, however, exhibits a three-dimensional nature. The reason is that the centrifugal force acting normal to the main direction of flow causes a secondary flow pattern in the pipe cross section

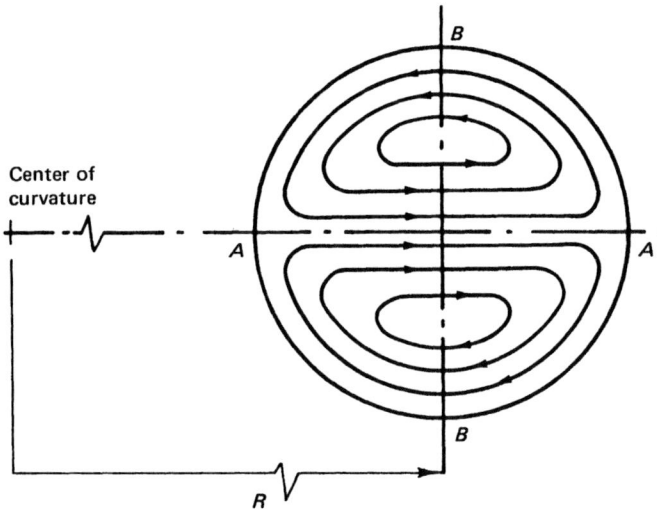

Figure 9.1 Secondary flow pattern in the cross section of a curved pipe [from Patankar, Pratap, and Spalding (1974)].

(Fig. 9.1). As a result, the point of maximum axial velocity shifts to the outside.

The developing laminar flow in a curved pipe was calculated by Patankar, Pratap, and Spalding (1974). A representative sample of the results is presented in Fig. 9.2, in which the axial velocity profiles on two different diameters are shown at successive locations along a bend, which is situated downstream of a straight section of the pipe. The velocity profile thus starts as a parabolic one and gradually distorts to its fully developed shape in the curved pipe. The computed results are compared with the experimental data of Austin (1971); the agreement can be seen to be quite good.

The paper presents many more results for flow and for heat transfer and compares them with experimental data. In a later study (Patankar, Pratap, and Spalding, 1975), the *turbulent* flow in curved pipes was computed by the use of a two-equation turbulence model.

9.2 COMBINED CONVECTION IN A HORIZONTAL TUBE

Patankar, Ramadhyani, and Sparrow (1978) have carried out a computational study of the fully developed laminar flow and heat transfer in a horizontal tube that is subjected to nonuniform circumferential heating. Two heating conditions, which are in evidence in the insets of Fig. 9.3, were considered. In one, the tube was uniformly heated over the top half and insulated over the

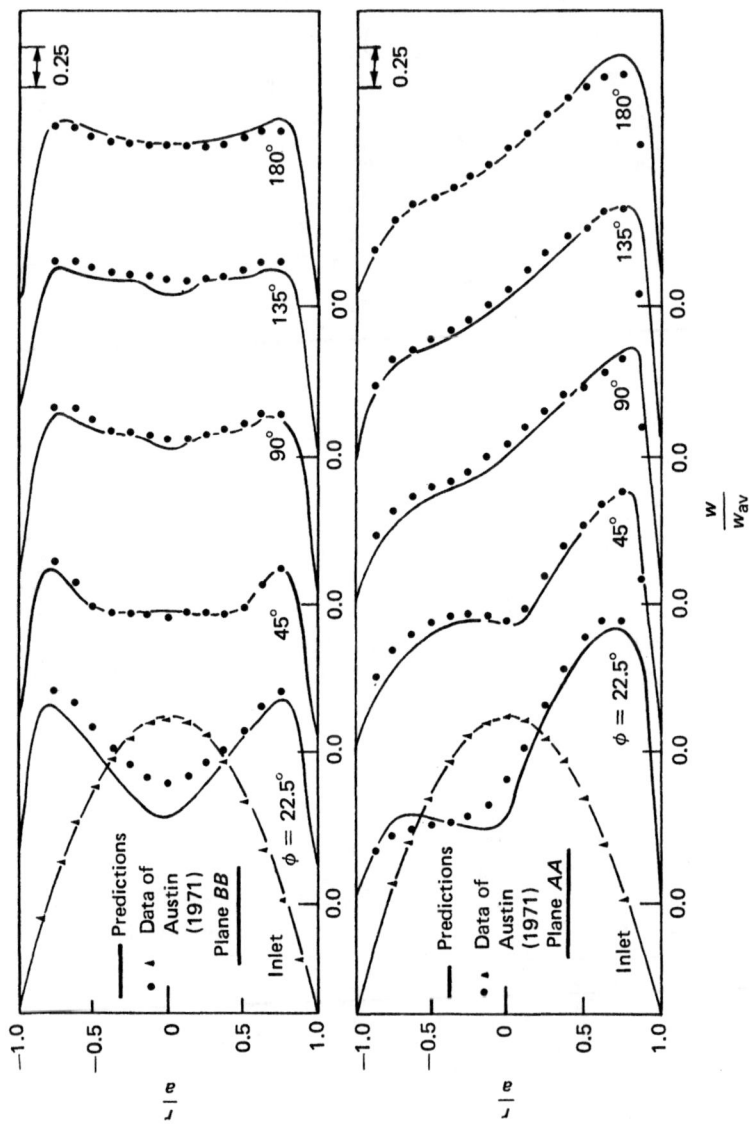

Figure 9.2 Development of axial velocity for the parameters $K = 372$ and $R/a = 29.1$ [from Patankar, Pratap, and Spalding (1974)].

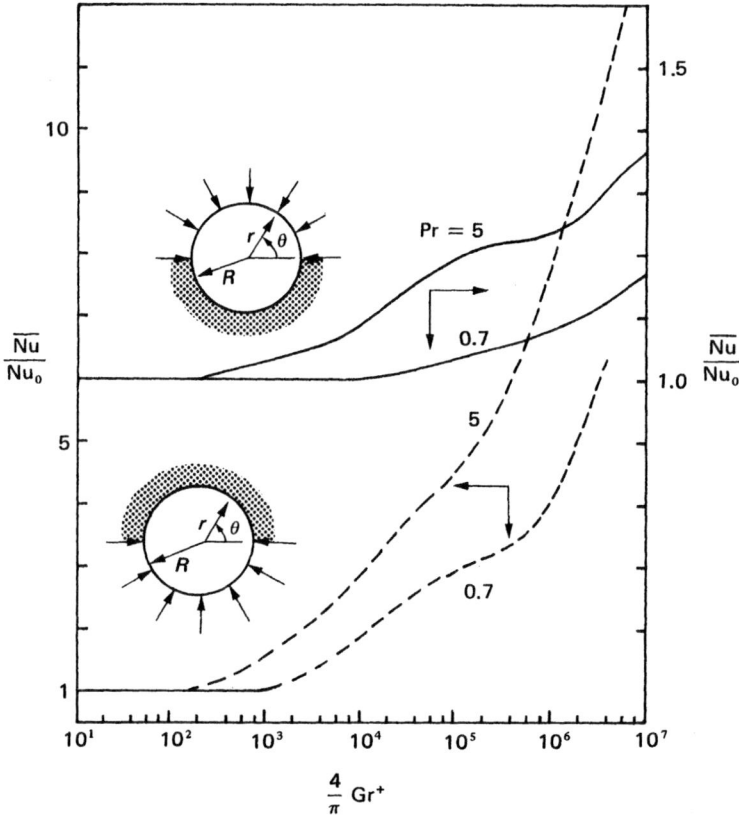

Figure 9.3 Average Nusselt numbers for the horizontal tube with nonuniform heating [from Patankar, Ramadhyani, and Sparrow (1978)].

bottom; in the other, the heated and insulated portions were reversed.

The nonuniform heating gives rise to a buoyancy-induced secondary flow, which leads to significantly higher Nusselt numbers than those for pure forced convection. The effect is particularly pronounced for the bottom-heating case and for the larger Prandtl number, as shown by the average Nusselt numbers plotted in Fig. 9.3. The abscissa is a multiple of the modified Grashof number Gr^+.

Further insight into these results can be obtained from the isotherms and streamlines over the tube cross section. The results for bottom heating are presented in Fig. 9.4 for three different values of Gr^+. In each cross-sectional representation, the isotherms are plotted on the left, and the streamlines on the right. The secondary flow caused by the nonuniform heating can be clearly seen. At the highest Grashof number, the streamline pattern is rather complicated, there being a tendency to form a "thermal" above the lowest

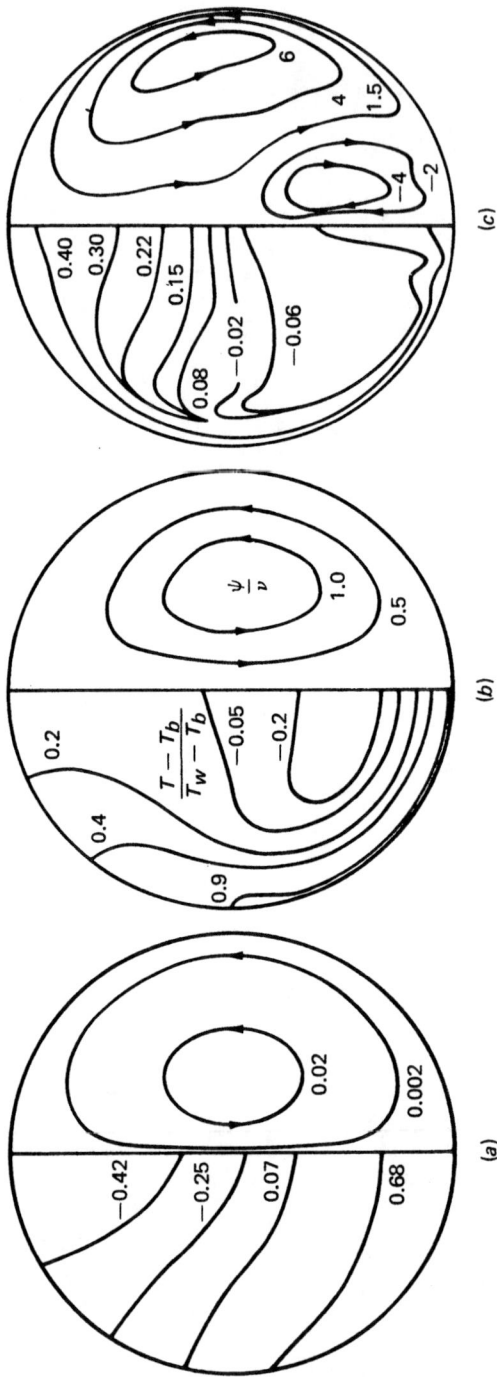

Figure 9.4 Isotherms and streamline maps for bottom heating; Pr = 5, and the values of $(4/\pi)Gr^{+}$ are (a) 10, (b) 10^4, and (c) 0.5×10^7 [from Patankar, Ramadhyani, and Sparrow (1978)].

point in the cross section. The isotherms for this case exhibit a kind of stably stratified structure in the top half, tend to follow the contour of the tube in the bottom half, and indicate the rising thermal at the very bottom of the tube.

9.3 MELTING AROUND A VERTICAL PIPE

We shall now consider the situation shown in Fig. 9.5. A vertical pipe carrying a hot fluid is embedded in a solid that is at its fusion temperature. With only conduction heat transfer acting at the beginning, the melt layer has a uniform thickness. But natural convection soon becomes influential and causes the fluid at the top to be hotter than that at the bottom. This results in the inclined interface as shown, with the largest thickness of the melt layer at the top.

A numerical solution for the situation described was obtained by Sparrow, Patankar, and Ramadhyani (1977). A grid in a transformed co-ordinate system was employed, which always fitted the everchanging and irregular shape of the melt region. In the unsteady solution, the interface was regarded as temporarily stationary during each time step; its position was readjusted before starting the next time step to account for the interface heat transfer.

The time-dependent variation of the heat transfer rate at the pipe surface is shown in Fig. 9.6. For our purposes here, it is best to ignore the various parameters in the figure and concentrate on the trends. At early times, the situation is governed by conduction, which causes a decrease in heat transfer as the increasing thickness of the melt layer offers a greater resistance. This is

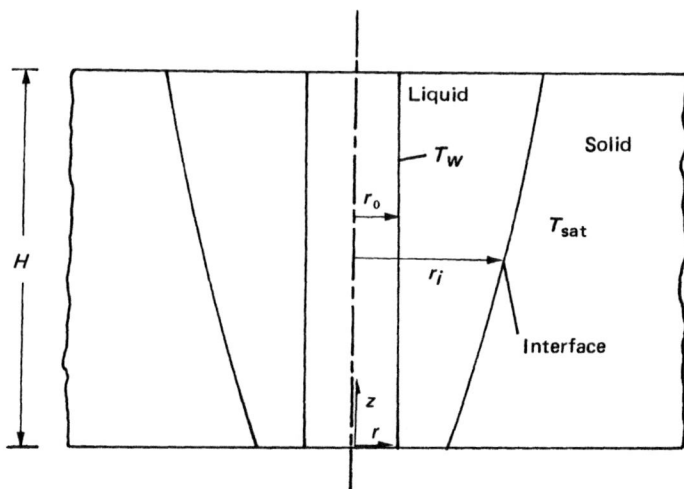

Figure 9.5 Melting problem [from Sparrow, Patankar, and Ramadhyani (1977)].

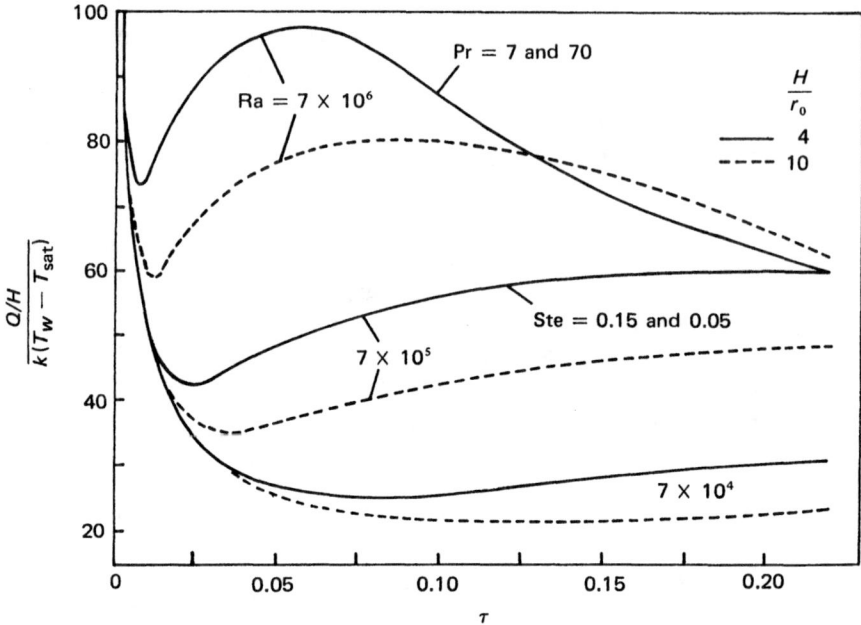

Figure 9.6 Timewise variation of the pipe heat transfer rate [from Sparrow, Patankar, and Ramadhyani (1977)].

followed by an increase in the heat transfer rate that is brought about by the action of natural convection. At large times, the rate of heat transfer is seen to decrease again; by now, the melt region is so large that heat is carried only by the recirculating flow, which itself experiences growing resistance along the top wall.

The natural convection in the melt region and the shape of the interface can be seen in Fig. 9.7 for three representative cases, for which the streamline patterns are shown. For the early-time case, the conduction-dominated melt region is nearly rectangular. The two other cases show the typical velocity patterns and interface shapes that result from significant natural convection.

9.4 TURBULENT FLOW AND HEAT TRANSFER IN INTERNALLY FINNED TUBES

A circular tube with longitudinal internal fins is considered to be an effective device for heat transfer enhancement. The fully developed flow and heat transfer in such a tube were computed by the use of a mixing-length model formulated for the cross-sectional geometry shown in Fig. 9.8. Complete details of the model and the resulting solutions are given in Patankar,

Ivanović, and Sparrow (1979). It is sufficient to note here that the model calculates the local mixing length based on the distances of a point from both the fin surface and the tube wall, and that the turbulent viscosity is influenced by the velocity gradients in both the radial and circumferential directions. The model incorporates a single adjustable constant, which was

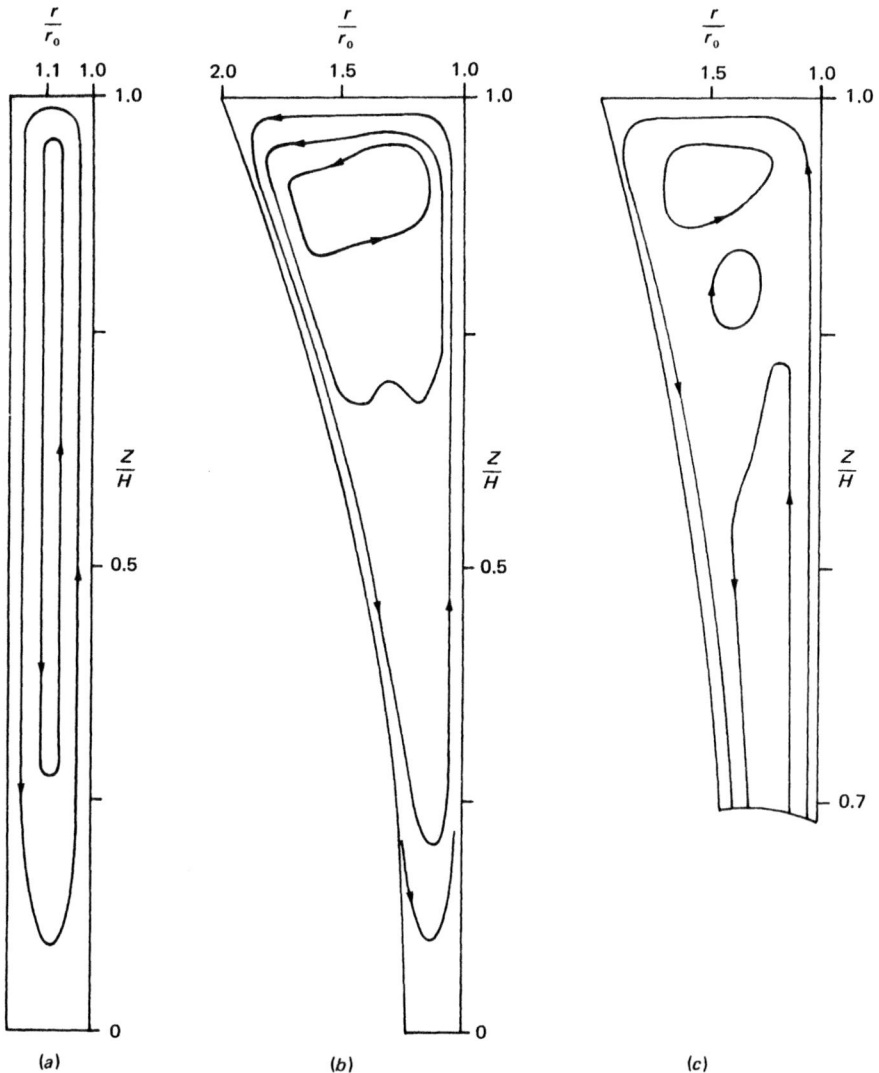

Figure 9.7 Representative flow patterns. The early situation is shown by (a), while (b) and (c) result from vigorous natural convection [from Sparrow, Patankar, and Ramadhyani (1977)].

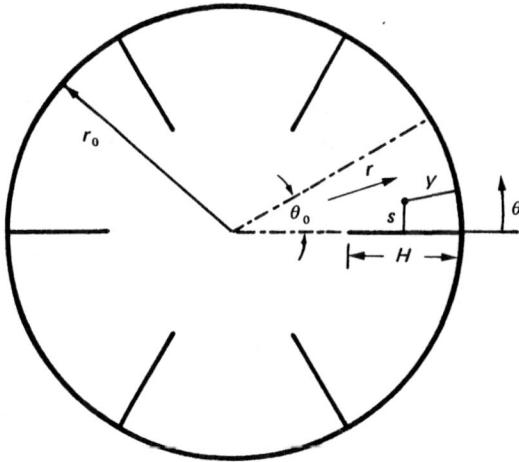

Figure 9.8 Cross-sectional geometry of an internally finned tube [from Patankar, Ivanović, and Sparrow (1979)].

chosen to give good agreement with the experimental data for air flow reported by Carnavos (1977).

Figure 9.9 shows the comparison of the predicted values of the Nusselt number and the friction factor with experimental data. In a way, the satisfactory agreement shown is not surprising, because the adjustable constant in the model was derived from the same experimental data. On the other hand, that the adjustment of a *single* constant is able to give good predictions for both

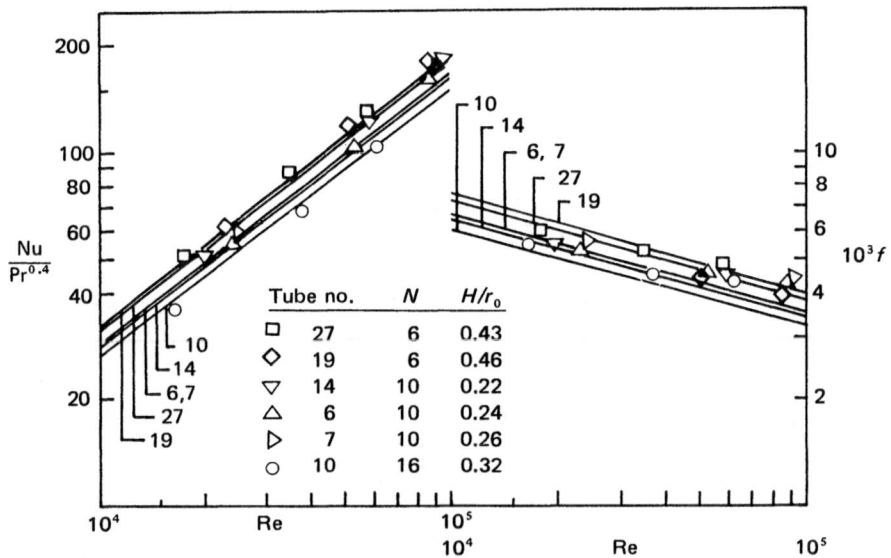

Tube no.	N	H/r_0
□ 27	6	0.43
◇ 19	6	0.46
▽ 14	10	0.22
△ 6	10	0.24
▷ 7	10	0.26
○ 10	16	0.32

Figure 9.9 Comparison of predicted values of the Nusselt number and friction factor with the experimental data of Carnavos (1977) [from Patankar, Ivanović, and Sparrow (1979)].

Nu and f over a range of Reynolds numbers and for different numbers and heights of fins is a significant achievement of the model.

9.5 A DEFLECTED TURBULENT JET

A turbulent jet issuing from a circular orifice can be analyzed as a two-dimensional parabolic flow. However, when the jet is deflected by a stream normal to its axis, an interesting three-dimensional elliptic situation arises, as schematically shown in Fig. 9.10. Chimney plumes, flow under a V/STOL aircraft, and some film-cooling situations involve the deflected-jet configuration.

Patankar, Basu, and Alpay (1977) obtained a numerical solution for the three-dimensional velocity field of the deflected jet on the basis of the k-ϵ model of turbulence. Thus, in addition to the momentum and continuity equations, two differential equations for the turbulence quantities, namely the turbulence kinetic energy k and its dissipation rate ϵ, were solved. The standard values of the empirical constants in the k-ϵ model, as recommended by Launder and Spalding (1974), were used; they were not adjusted to procure better agreement with experimental data.

The predicted position of the jet center line is shown in Fig. 9.11 for various ratios of the jet velocity to the mainstream velocity. Also shown are

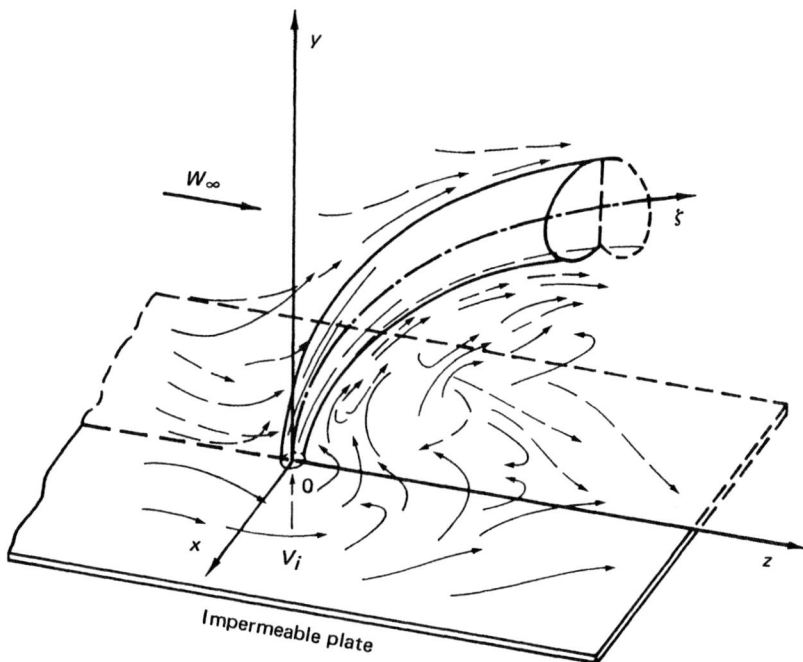

Figure 9.10 Deflected-jet situation [from Patankar, Basu, and Alpay (1977)].

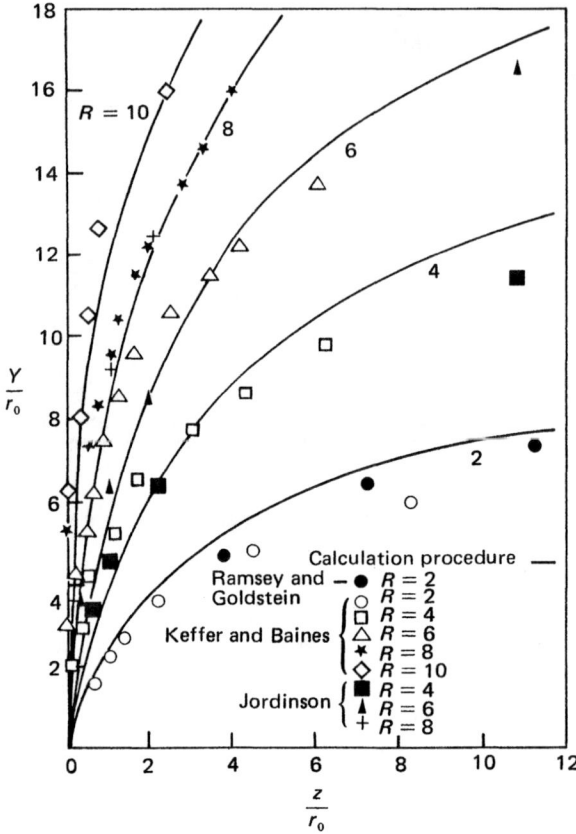

Figure 9.11 Position of the jet centerline for different jet-to-mainstream velocity ratios [from Patankar, Basu, and Alpay (1977)].

the experimental data of Ramsey and Goldstein (1970), Keffer and Baines (1963), and Jordinson (1958). Within the experimental scatter, the agreement of the numerical predictions with the data can be judged as satisfactory.

In Fig. 9.12, we compare the computed velocity profiles with the measured ones from Ramsey and Goldstein (1970). These are the profiles of the z-direction velocity along the central yz plane for four different values of z; the ratio of the jet velocity to the mainstream velocity is 2. Again, the agreement is reasonable.

9.6 A HYPERMIXING JET WITHIN A THRUST–AUGMENTING EJECTOR

A thrust-augmenting ejector is an arrangement for increasing the thrust of a primary jet by entraining secondary air from the atmosphere. It has possible

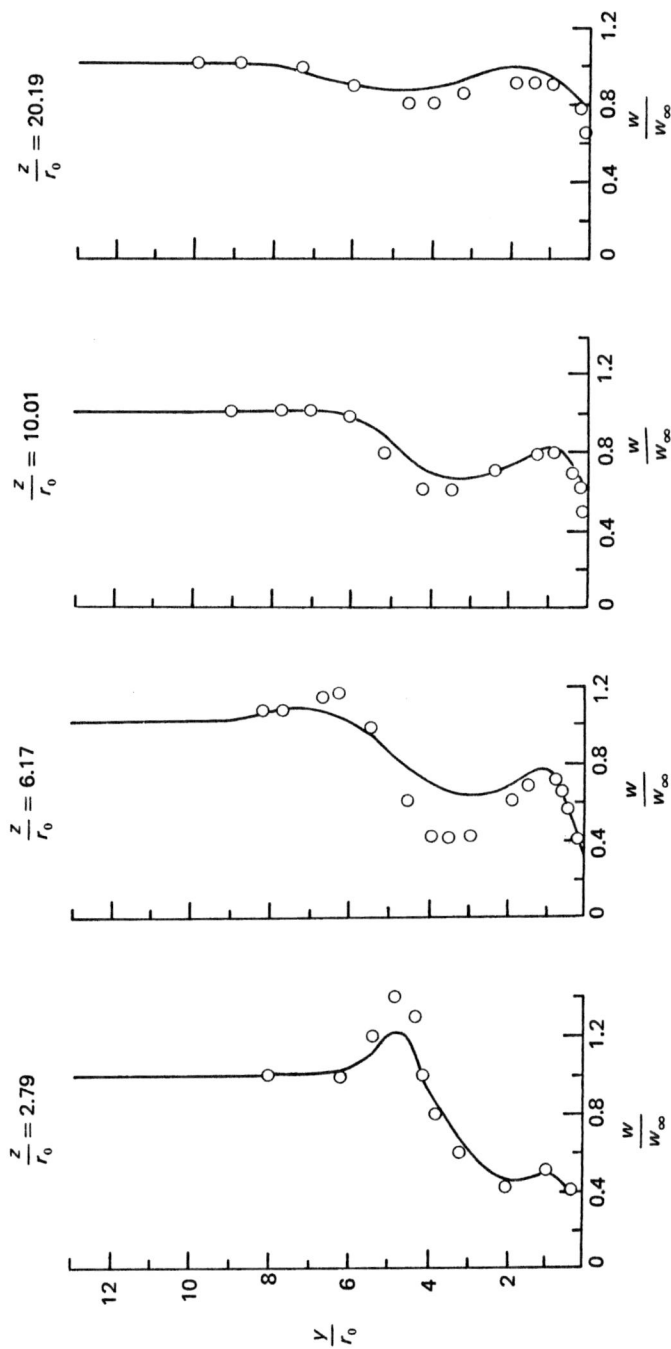

Figure 9.12 Profiles of the z-direction velocity [from Patankar, Basu, and Alpay (1977)].

applications in V/STOL aircraft. Ordinary jets from slot nozzles require long mixing ducts to produce any significant thrust augmentation. Since such long ducts are unsuitable for practical aircraft applications, a hypermixing nozzle is used to accelerate the mixing process. Here we summarize the computational investigation of a hypermixing-jet ejector reported by DeJoode and Patankar (1978).

The geometry of the hypermixing nozzle and the resulting flow field are shown in Fig. 9.13. The nozzle exit is divided into several segments. The flow issuing from these segments is given an upward or downward velocity component in an alternating fashion; this is shown schematically in the inset of Fig. 9.13. These alternate velocity components lead to the formation of streamwise vortices, indicated by the arrows on a cross-stream plane in the figure. Also shown are the profiles of the velocity in the main flow direction. The velocity maxima in front of adjacent segments can be seen to lie respectively above and below the center line, while in front of the dividing line between two segments the velocity profile has two peaks.

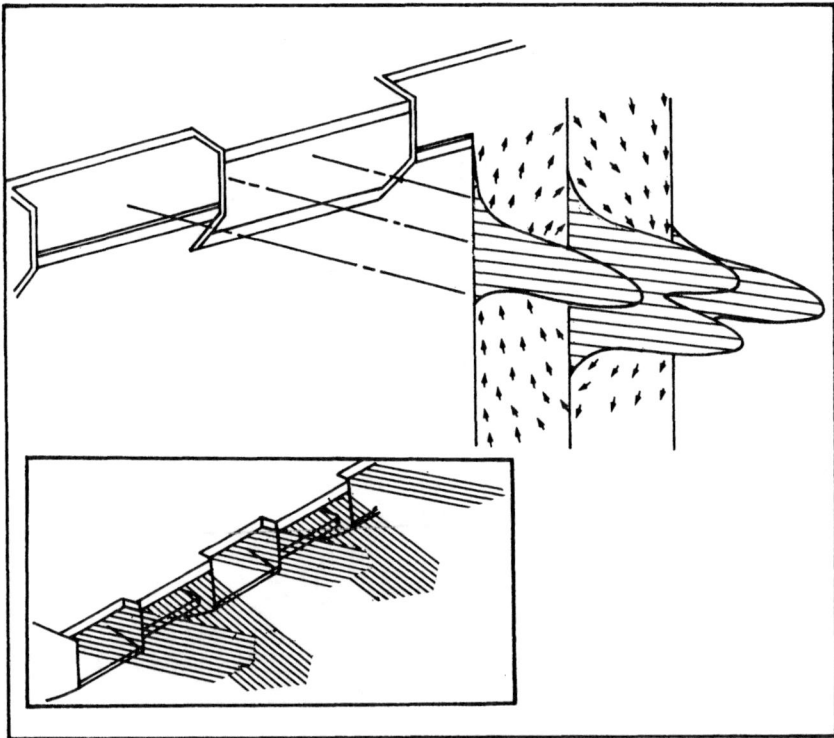

Figure 9.13 Geometry and the flow field of a hypermixing jet [from DeJoode and Patankar (1978)].

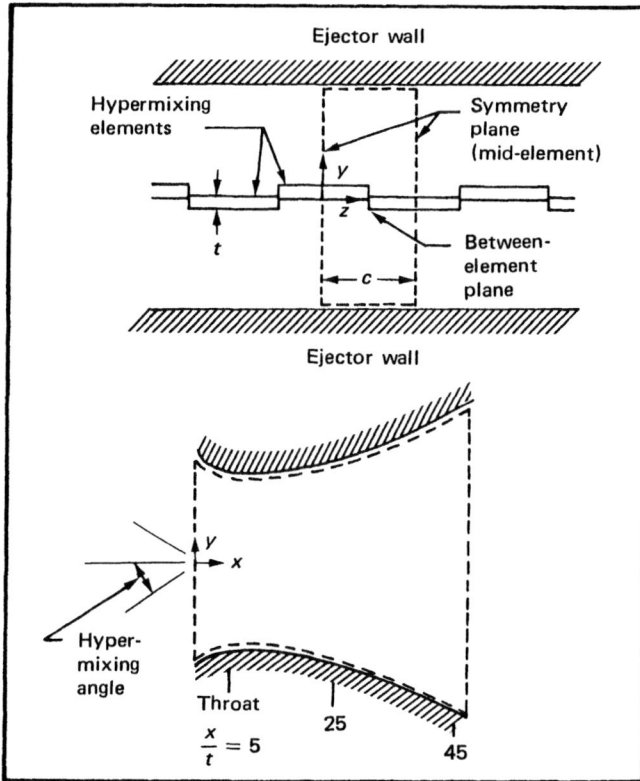

Figure 9.14 Configuration chosen for numerical prediction [from DeJoode and Patankar (1978)].

For the situation chosen for computer analysis, the hypermixing jet was placed in a diffuser as shown in Fig. 9.14. The computation was performed by a marching procedure for the three-dimensional parabolic flow. The k-ϵ model of turbulence, with the standard values of constants from Launder and Spalding (1974), was used.

The comparison of predicted and measured velocity profiles is shown in Fig. 9.15. All the qualitative features of the flow field—such as the double peak between the elements, the appearance of a second peak at the mid-element location, and the merging of the two peaks at a far-downstream position—are correctly predicted; the quantitative agreement is also fairly good.

The pressure rise through the diffuser is considered as a convenient measure of the thrust augmentation achieved. The predicted pressure rise through the ejector is compared with the experimental data in Fig. 9.16. Once again, the agreement can be regarded as reasonable.

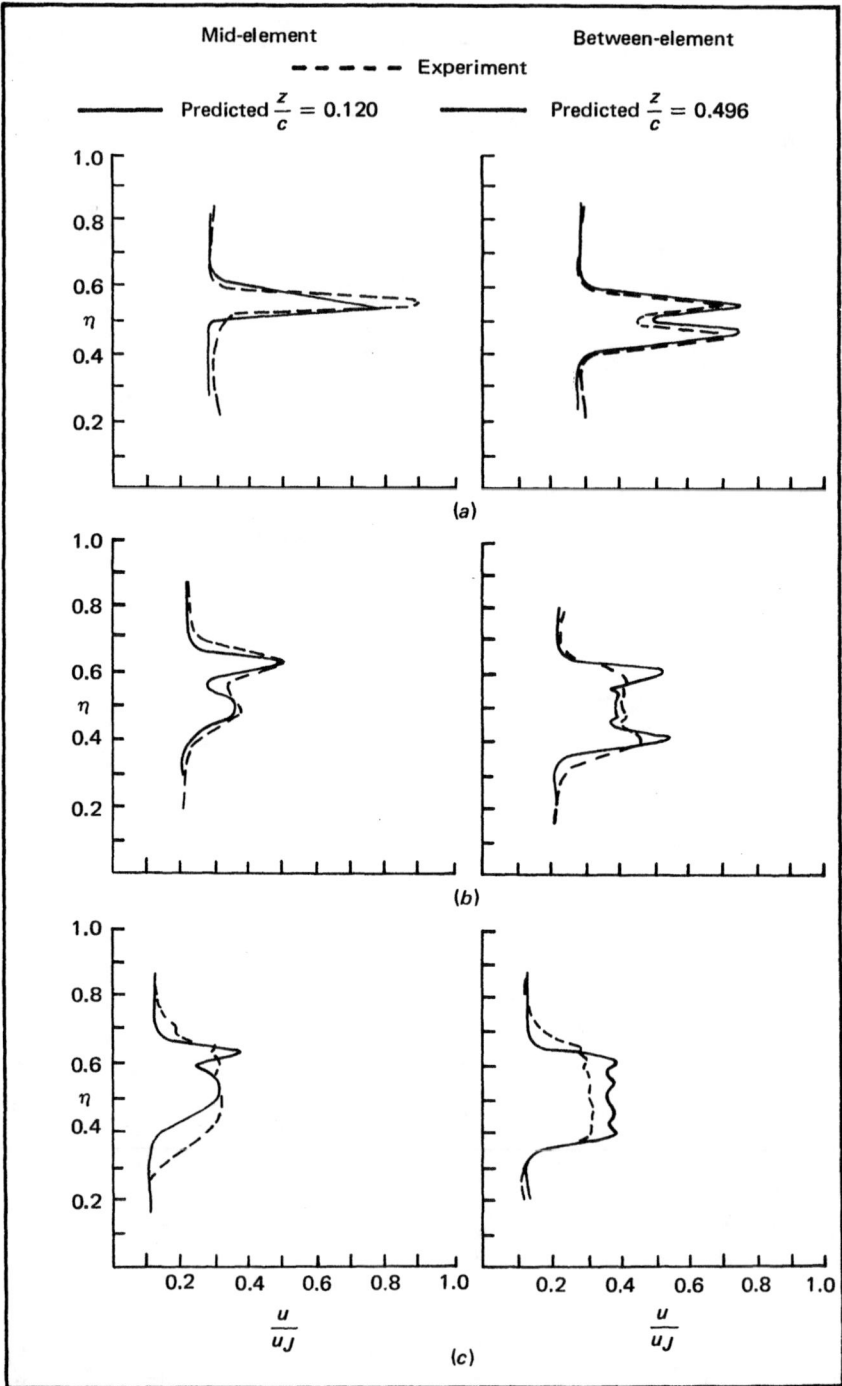

Figure 9.15 Comparison of predicted and measured velocity profiles. (*a*) $x/t = 5$; (*b*) $x/t = 25$; (*c*) $x/t = 45$ [from DeJoode and Patankar (1978)].

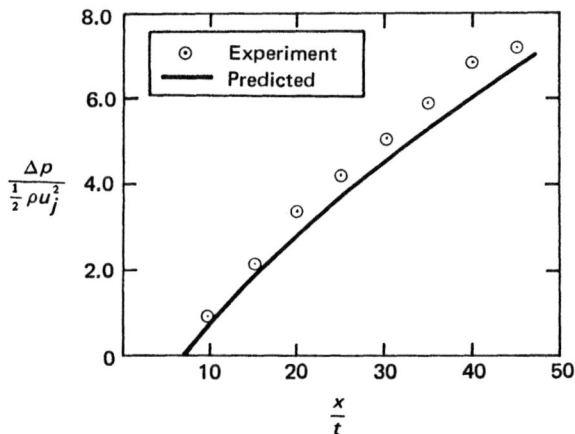

Figure 9.16 Pressure rise in the diffuser [from DeJoode and Patankar (1978)].

9.7 A PERIODIC FULLY DEVELOPED DUCT FLOW

Let us now consider the calculation of the fluid flow and heat transfer for the configuration shown in Fig. 9.17. The situation is characterized by the repetition of an identical geometrical module, such as the transverse plates shown. Such configurations are common in heat exchangers and in heat transfer augmentation devices. It is easy to see that if the entire region, consisting of a large number of modules, were used as a calculation domain, the required computer storage and computer time would be truly excessive. A practical alternative is provided by recognizing that, beyond a certain development length, the velocity field will repeat itself module after module, and the temperature field also will exhibit a kind of similarity. It is, therefore, possible to calculate the flow and heat transfer directly for the typical module shown

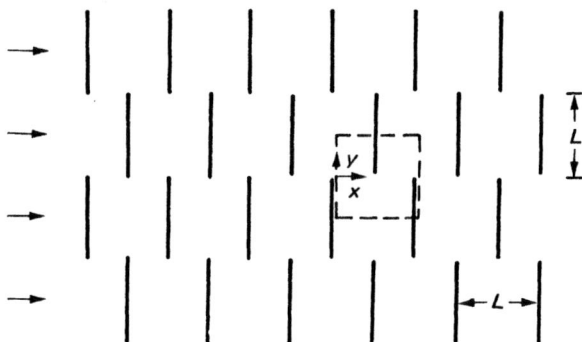

Figure 9.17 Transverse-plate array [from Patankar, Liu, and Sparrow (1977)].

by dashed lines in Fig. 9.17, without having to deal with the entrance-region problem.

The calculation for the module may at first sight appear to be burdened with a difficulty: We do not have known values of velocity, temperature, etc. at the upstream and downstream boundaries of the module. But further thought eliminates the difficulty. When the fluid leaves the module, it enters an identical next module. Therefore, the situation is conceptually the same as if the fluid leaving the module were (somehow) to reenter the *same* module at the upstream end. In this view, the upstream and downstream boundaries do not form boundary locations at all; all streamwise stations are as if arranged in an endless loop.

This conceptual framework is sufficient to formulate the numerical solution, which is described in detail by Patankar, Liu, and Sparrow (1977). A representative solution for the module shown in Fig. 9.17 is presented in Fig. 9.18 in the form of the streamlines. It can be noted that the flow has to take a rather tortuous path to get around the transverse plates. This leads to the large recirculation zone on the downstream side of each plate. The heat transfer calculation for the same situation with a Prandtl number of 0.7 leads to the Nusselt numbers plotted in Fig. 9.19. The higher Nusselt numbers on

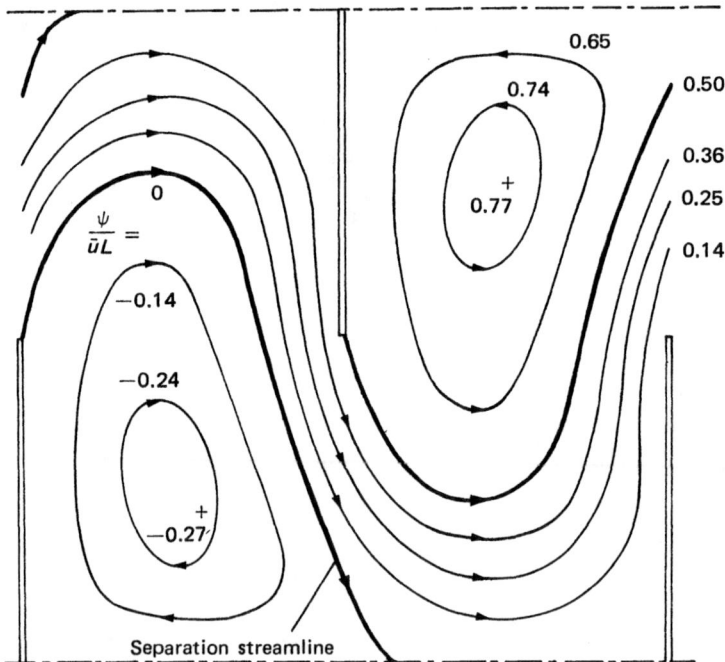

Figure 9.18 Predicted flow field for a Reynolds number of 1040 [from Patankar, Liu, and Sparrow (1977)].

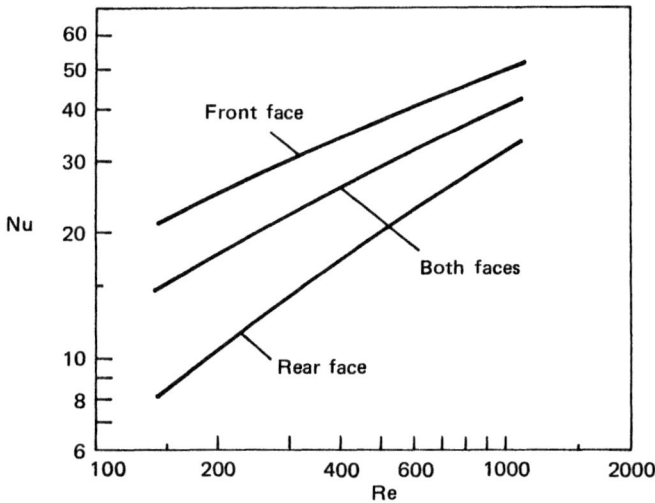

Figure 9.19 Predicted Nusselt numbers [from Patankar, Liu, and Sparrow (1977)].

the front face of the plates are caused by the impinging flow there, while the slow recirculation zone on the back face gives much lower values. The increase in the Nusselt number with the Reynolds number is, in itself, unlike the case of conventional thermally developed duct flows, for which the Nusselt number is independent of the Reynolds number.

9.8 THERMAL–HYDRAULIC ANALYSIS
OF A STEAM GENERATOR

This last example is included here to illustrate two main points: that useful computations for realistic large-scale industrial equipment can now be made, and that the "distributed-resistance" concept can be effectively employed to analyze configurations such as heat exchangers, steam generators, condensers, and cooling towers.

The distributed-resistance concept is applicable to cases in which a fluid flows through an enclosure that is filled with numerous solid objects such as rods, tubes, or slats. The situation is then treated much like flow in porous media, with distributed sinks of momentum and sources or sinks of heat produced by the solid objects. The distributed resistance can be obtained from detailed computations such as the one in Section 9.7 or directly from empirical correlations for the appropriate configuration.

The thermal-hydraulic analysis of a steam generator, which is described by Patankar and Spalding (1978), was carried out for the configuration shown in Fig. 9.20. The cylindrical shell is uniformly filled with tubes (which are not shown in Fig. 9.20). The hot tube fluid rises upward in one half of the steam generator, turns through the U bend at the top, and flows downward in the other half. An economizer is housed in the lower part of the generator for the purpose of bringing the feedwater up to the saturation temperature.

The numerical solution was carried out to obtain the three velocity components, the pressure, the enthalpy for the shell fluid, and the enthalpy (or temperature) of the tube fluid. For the situation considered, the tube fluid remained in the liquid phase throughout, and its mass flow rate was known from the inlet conditions.

The computed velocity field on the central vertical plane is shown in Fig. 9.21. The arrows denote the velocity vectors in both magnitude and direction. The general magnitude of the velocity can be seen to increase as the fluid rises in the steam generator; this is in response to the lower values of density in the upper part. The velocity field in the lower left-hand corner of the figure indicates the zig-zag flow path through the economizer.

Figure 9.22 shows the steam-quality distribution on the central vertical plane. The lower left-hand corner is blank because the fluid in the economizer is mostly subcooled water. In general, the qualities on the right side (i.e., the "hot" side) are greater than those on the left side. This disparity is seen to exist all the way to the exit.

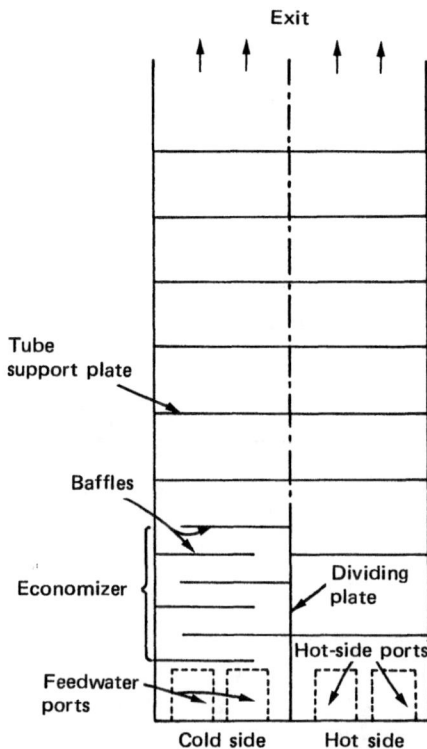

Figure 9.20 Steam-generator configuration. (The tubes are not shown; the figure is not drawn to scale; the horizontal dimension is shown enlarged by a factor of about 2.) [From Patankar and Spalding (1978).]

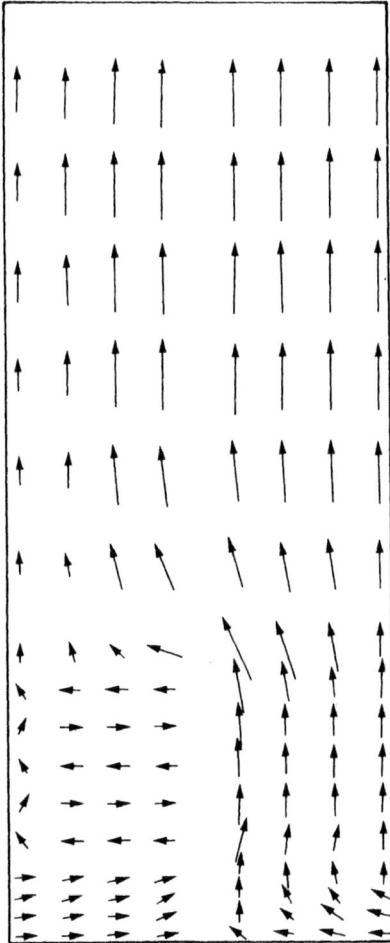

Figure 9.21 Shell-side velocity vectors on the vertical plane of symmetry [from Patankar and Spalding (1978)].

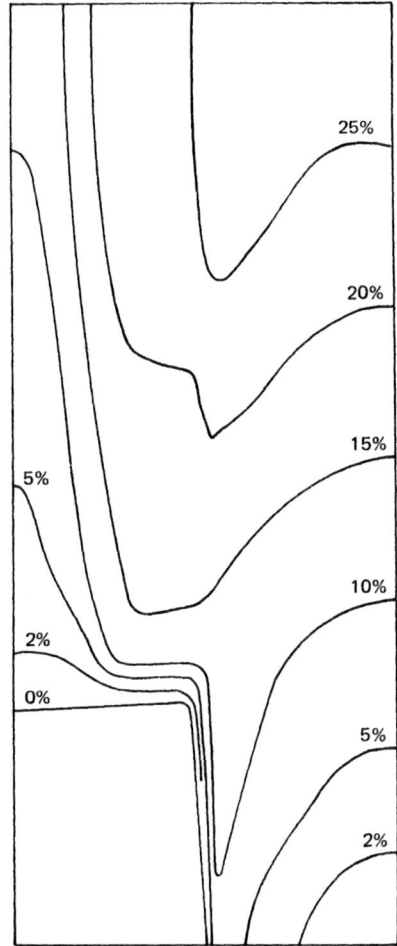

Figure 9.22 Contours of steam quality on the vertical plane of symmetry [from Patankar and Spalding (1978)].

9.9 CLOSING REMARKS

In this book, we have developed a numerical method for heat transfer, fluid flow, and related phenomena; evolved a philosophy of numerical computation through physical understanding and insight; and presented illustrative examples of actual computations. Sufficient details of the method are given to enable readers to write their own computer programs. The readers are also equipped with meaningful criteria with which to judge other methods. The purpose of the book would be well served if each reader became an active practitioner of, and possibly an innovator in, the exciting field of numerical heat transfer and fluid flow.

NOMENCLATURE

A	convection-diffusion coefficient, Eq. (5.37); also used to denote area in Chapter 6
a	coefficient in the discretization equation
B	convection-diffusion coefficient, Eq. (5.37)
B_x	x-direction body force, Eq. (2.11)
b	constant term in the discretization equation
c	specific heat
D	diffusion conductance, Eq. (5.9)
d	coefficient of the pressure-difference term, Eq. (6.16)
F	flow rate through a control-volume face, Eq. (5.9)
f	weighting factor, Eq. (4.34)
f_e	length ratio, Eq. (4.6)
G	generation rate of turbulence energy, Eq. (2.12)
h	specific enthalpy in Chapter 2; heat transfer coefficient in Chapter 4
i	inertia used for underrelaxation, Eq. (4.56)
J	total (convection + diffusion) flux
J_l	diffusion flux of chemical species l
J^*	normalized flux, Eq. (5.35)
k	thermal conductivity; also used to denote the turbulence kinetic energy, Eq. (2.12)
m_l	mass fraction of the chemical species l
P	Peclet number, Eq. (5.18); also used as a TDMA coefficient in Chapter 4
p	pressure
p'	pressure correction

Q	TDMA coefficient
q	heat flux
R	residual, Eqs. (3.8) and (7.5)
R_l	rate of generation of species l by chemical reaction, Eq. (2.2)
r	radial coordinate
S	general source term, Eq. (2.13)
S_C	constant part of the linearized source term, Eq. (3.16)
S_h	volumetric rate of heat generation, Eq. (2.5)
S_P	coefficient of T_P (or ϕ_P) in the linearized source expression, Eq. (3.16)
T	temperature
t	time
u	x-direction velocity
$\mathbf{u}$	velocity vector
$\hat{u}$	pseudovelocity in the x direction, Eq. (6.26)
u^*	velocity based on the guessed pressure p^*
V_x	viscous source term in Eq. (2.11)
v	y-direction velocity
$\hat{v}, v^*$	similar to $\hat{u}, u^*$
W	weighting function, Eq. (3.9)
w	z-direction velocity
$\hat{w}, w^*$	similar to $\hat{u}, u^*$
x, y, z	coordinates
α	relaxation factor, Eq. (4.55)
α_p	relaxation factor for pressure, Eq. (6.24)
Γ	general diffusion coefficient, Eq. (2.13)
Γ_l	diffusion coefficient for species l, Eq. (2.3)
Δt	time step
Δx	x-direction width of the control volume
δx	x-direction distance between two adjacent grid points
$\Delta y, \delta y$	similar to $\Delta x, \delta x$
$\Delta z, \delta z$	similar to $\Delta x, \delta x$
ϵ	turbulence dissipation rate, Eq. (2.12)
μ	viscosity
ρ	density
ϕ	general dependent variable, Eq. (2.13)

Subscripts

B	neighbor in the negative z direction, i.e., at the bottom
b	control-volume face between P and B
E	neighbor in the positive x direction, i.e., on the east side
e	control-volume face between P and E
N	neighbor in the positive y direction, i.e., on the north side

n	control-volume face between P and N
nb	general neighbor grid point
P	central grid point under consideration
S	neighbor in the negative y direction, i.e., on the south side
s	control-volume face between P and S
T	neighbor in the positive z direction, i.c,, at the top
t	control-volume face between P and T
W	neighbor in the negative x direction, i.e., on the west side
w	control-volume face between P and W

Superscripts

1	new value (at time $t + \Delta t$) of the dependent variable
0	old value (at time t) of the variable
*	previous-iteration value of a variable; also velocities based on a guessed pressure

Special symbol

$[\![A, B, C, \ldots]\!]$ largest of A, B, C, . . .

REFERENCES

Abdel-Wahed, R. M., Patankar, S. V., and Sparrow, E. M. (1976). Fully Developed Laminar Flow and Heat Transfer in a Square Duct with One Moving Wall, *Lett. Heat Mass Transfer*, vol. 3, p. 355.

Austin, L. R. (1971). The Development of Viscous Flow within Helical Coils, Ph.D. thesis, University of Utah, Salt Lake City.

Aziz, K. and Hellums, J. D. (1967). Numerical Solution of the Three-Dimensional Equations of Motion for Laminar Natural Convection, *Phys. Fluids*, vol. 10, p. 314.

Baliga, B. R. and Patankar, S. V. (1979a). A New Finite-Element Formulation for Convection-Diffusion Problems, to be published.

Baliga, B. R. and Patankar, S. V. (1979b). A Control-Volume-Based Finite-Element Method for Two-Dimensional Flows, to be published.

Barakat, H. Z. and Clark, J. A. (1966). Analytical and Experimental Study of Transient Laminar Natural Convection Flows in Partially Filled Containers, *Proc. 3d Int. Heat Transfer Conf.*, Chicago, vol. II, paper 57, p. 152.

Bergeles, G., Gosman, A. D., and Launder, B. E. (1976). The Prediction of Three-Dimensional Discrete-Hole Cooling Processes, part 1, Laminar Flow, *J. Heat Transfer*, vol. 98, p. 379.

Bergeles, G., Gosman, A. D., and Launder, B. E. (1978). The Turbulent Jet in a Cross Stream at Low Injection Rates: A Three-Dimensional Numerical Treatment, *Num. Heat Transfer*, vol. 1, p. 217.

Caretto, L. S., Curr, R. M., and Spalding, D. B. (1972). Two Numerical Methods for Three-Dimensional Boundary Layers, *Comp. Methods Appl. Mech. Eng.*, vol. 1, p. 39.

Caretto, L. S., Gosman, A. D., Patankar, S. V., and Spalding, D. B. (1972). Two Calculation Procedures for Steady, Three-Dimensional Flows with Recirculation, *Proc. 3d Int. Conf. Num. Methods Fluid Dyn.*, Paris, vol. II, p. 60.

Carnavos, T. C. (1977). Cooling Air in Turbulent Flow with Internally Finned Tubes, AIChE paper 4, *17th Natl. Heat Transfer Conf.*

Courant, R., Isaacson, E., and Rees, M. (1952). On the Solution of Non-Linear

Hyperbolic Differential Equations by Finite Differences, *Comm. Pure Appl. Math.*, vol. 5, p. 243.

Crank, J. and Crowley, A. B. (1978). Isotherm Migration along Orthogonal Flow Lines in Two Dimensions, *Int. J. Heat Mass Transfer*, vol. 21, p. 393.

Crank, J. and Gupta, R. S. (1975). Isotherm Migration Method in Two Dimensions, *Int. J. Heat Mass Transfer*, vol. 18, p. 1101.

Crank, J. and Phahle, R. D. (1973). Melting Ice by the Isotherm Migration Method, *Bull. Inst. Math. Appl.*, vol. 9, p. 12.

DeJoode, A. D. and Patankar, S. V. (1978). Prediction of Three-Dimensional Turbulent Mixing in an Ejector, *AIAA J.*, vol. 16, p. 145.

de Vahl Davis, G. and Mallinson, G. D. (1972). False Diffusion in Numerical Fluid Mechanics, Univ. of New South Wales, School of Mech. and Ind. Eng. Rept. 1972/FMT/1.

Dix, D. M. (1963). The Magnetohydrodynamic Flow past a Non-Conducting Flat Plate in the Presence of a Transverse Magnetic Field, *J. Fluid Mech.*, vol. 15, p. 449.

Dix, R. C. and Cizek, J. (1970). The Isotherm Migration Method for Transient Heat Conduction Analysis, *Proc. 4th Int. Heat Transfer Conf.*, Paris, vol. 1, p. Cu1.1.

Durst, F. and Rastogi, A. K. (1977). Theoretical and Experimental Investigations of Turbulent Flows, *Symp. on Turbulent Shear Flows*, Pennsylvania State Univ., vol. 1, p. 18.1.

Finlayson, B. A. (1972). *The Method of Weighted Residuals and Variational Principles*, Academic, New York.

Fromm, J. E. and Harlow, F. H. (1963). Numerical Solution of the Problem of Vortex Street Development, *Phys. Fluids*, vol. 6, p. 975.

Ganesan, V., Spalding, D. B., and Murthy, B. S. (1978). Experimental and Theoretical Investigation of Flow behind an Axi-Symmetrical Baffle in a Circular Duct, *J. Inst. Fuel*, vol. 51, p. 144.

Gentry, R. A., Martin, R. E., and Daly, B. J. (1966). An Eulerian Differencing Method for Unsteady Compressible Flow Problems, *J. Comp. Phys.*, vol. 1, p. 87.

Gosman, A. D., Pun, W. M., Runchal, A. K., Spalding, D. B., and Wolfshtein, M. (1969). *Heat and Mass Transfer in Recirculating Flows*, Academic, New York.

Harlow, F. H. and Welch, J. E. (1965). Numerical Calculation of Time-Dependent Viscous Incompressible Flow of Fluid with Free Surface, *Phys. Fluids*, vol. 8, p. 2182.

Issa, R. I. and Lockwood, F. C. (1977). Prediction of Two-Dimensional Supersonic Viscous Interactions near Walls, *AIAA J.*, vol. 15, p. 182.

Jordinson, R. (1958). Flow in a Jet Directed Normal to the Wind, Aero. Res. Council R&M 3074.

Keffer, J. F. and Baines, W. D. (1963). The Round Turbulent Jet in a Cross Wind, *J. Fluid Mech.*, vol. 15, p. 481.

Khalil, E. E., Spalding, D. B., and Whitelaw, J. H. (1975). The Calculation of Local Flow Properties in Two-Dimensional Furnaces, *Int. J. Heat Mass Transfer*, vol. 18, p. 775.

King, H. H. (1976). A Poisson Equation Solver for Rectangular or Annular Regions, *Int. J. Num. Methods Eng.*, vol. 10, p. 799.

Kolmogorov, A. N. (1942). Equations of Turbulent Motion in an Incompressible Fluid, *Izv. Akad. Nauk SSSR Ser. Fiz.*, vol. 1-2, p. 56 (English translation: Imperial College, Mech. Eng. Dept. Rept. ON/6, 1968).

Launder, B. E. and Spalding, D. B. (1972). *Mathematical Models of Turbulence*, Academic, New York.

Launder, B. E. and Spalding, D. B. (1974). The Numerical Computation of Turbulent Flow, *Comp. Methods Appl. Mech. Eng.*, vol. 3, p. 269.

Lilley, D. G. (1976). Primitive Pressure-Velocity Code for the Computation of Strongly Swirling Flows, *AIAA J.*, vol. 14, p. 749.

Majumdar, A. K., Pratap, V. S., and Spalding, D. B. (1977). Numerical Computation of Flow in Rotating Ducts, *J. Fluids Eng.*, vol. 99, p. 148.

Majumdar, A. K. and Spalding, D. B. (1977). Numerical Computation of Taylor Vortices, *J. Fluid Mech.*, vol. 81, p. 295.

McGuirk, J. J. and Rodi, W. (1977). The Calculation of Three-Dimensional Turbulent Free Jets, Symp. on Turbulent Shear Flows, Pennsylvania State Univ., vol, 1, p. 1.29.

McGuirk, J. J. and Rodi, W. (1978). A Depth-Averaged Mathematical Model for the Near Field of Side Discharges into Open-Channel Flow, *J. Fluid Mech.*, vol. 86, p. 761.

Moon, L. F. and Rudinger, G. (1977). Velocity Distribution in an Abruptly Expanding Circular Duct, *J. Fluids Eng.*, vol. 99, p. 226.

Patankar, S. V. (1975). Numerical Prediction of Three-Dimensional Flows, in B. E. Launder (ed.), *Studies in Convection: Theory, Measurement and Applications*, vol. 1, Academic, New York.

Patankar, S. V. (1978). A Numerical Method for Conduction in Composite Materials, Flow in Irregular Geometries and Conjugate Heat Transfer, *Proc. 6th Int. Heat Transfer Conf.*, Toronto, vol. 3, p. 297.

Patankar, S. V. (1979a). A Calculation Procedure for Two-Dimensional Elliptic Situations, *Num. Heat Transfer*, vol. 2 (to be published).

Patankar, S. V. (1979b). The Concept of a Fully Developed Regime in Unsteady Heat Conduction, in J. P. Hartnett, T. F. Irvine, Jr., E. Pfender, and E. M. Sparrow (eds.), *Studies in Heat Transfer*, Hemisphere, Washington, D.C.

Patankar, S. V. and Baliga, B. R. (1978). A New Finite-Difference Scheme for Parabolic Differential Equations, *Num. Heat Transfer*, vol. 1, p. 27.

Patankar, S. V., Basu, D. K., and Alpay, S. A. (1977). Prediction of the Three-Dimensional Velocity Field of a Deflected Turbulent Jet, *J. Fluids Eng.*, vol. 99, p. 758.

Patankar, S. V., Ivanović, M., and Sparrow, E. M. (1979). Analysis of Turbulent Flow and Heat Transfer in Internally Finned Tubes and Annuli, *J. Heat Transfer*, vol. 101, p. 29.

Patankar, S. V., Liu, C. H., and Sparrow, E. M. (1977). Fully Developed Flow and Heat Transfer in Ducts Having Streamwise-Periodic Variations of Cross-Sectional Area, *J. Heat Transfer*, vol. 99, p. 180.

Patankar, S. V., Pratap, V. S., and Spalding, D. B. (1974). Prediction of Laminar Flow and Heat Transfer in Helically Coiled Pipes, *J. Fluid Mech.*, vol. 62, p. 539.

Patankar, S. V., Pratap, V. S., and Spalding, D. B. (1975). Prediction of Turbulent Flow in Curved Pipes, *J. Fluid Mech.*, vol. 67, p. 583.

Patankar, S. V., Rafiinejad, D., and Spalding, D. B. (1975). Calculations of the Three-Dimensional Boundary Layer with Solutions of All Three Momentum Equations, *Comp. Methods Appl. Mech. Eng.*, vol. 6, p. 283.

Patankar, S. V., Ramadhyani, S., and Sparrow, E. M. (1978). Effect of Circumferentially Nonuniform Heating in Laminar Combined Convection in a Horizontal Tube, *J. Heat Transfer*, vol. 100, p. 63. (Also see the Erratum in *J. Heat Transfer*, vol. 100, p. 367, 1978.)

Patankar, S. V., Rastogi, A. K., and Whitelaw, J. H. (1973). The Effectiveness of Three-Dimensional Film-Cooling Slots—Predictions, *Int. J. Heat Mass Transfer*, vol. 16, p. 1673.

Patankar, S. V. and Spalding, D. B. (1970). *Heat and Mass Transfer in Boundary Layers*, 2d ed., Intertext, London.

Patankar, S. V. and Spalding, D. B. (1972a). A Calculation Procedure for Heat, Mass and Momentum Transfer in Three-Dimensional Parabolic Flows, *Int. J. Heat Mass Transfer*, vol. 15, p. 1787.

Patankar, S. V. and Spalding, D. B. (1972b). A Computer Model for Three-Dimensional

Flow in Furnaces, *Proc. 14th Symp. (Int.) on Combustion*, The Combustion Inst., p. 605.

Patankar, S. V. and Spalding, D. B. (1974a). A Calculation Procedure for the Transient and Steady-State Behavior of Shell-and-Tube Heat Exchangers, chap. 7 in *Heat Exchangers: Design and Theory Sourcebook*, Hemisphere, Washington, D.C.

Patankar, S. V. and Spalding, D. B. (1974b). Simultaneous Predictions of Flow Pattern and Radiation for Three-Dimensional Flames, chap. 4 in *Heat Transfer in Flames*, Hemisphere, Washington, D.C.

Patankar, S. V. and Spalding, D. B. (1978). Computer Analysis of the Three-Dimensional Flow and Heat Transfer in a Steam Generator, *Forsch. Ingenieurwes.*, vol. 44, p. 47.

Patankar, S. V., Sparrow, E. M., and Ivanović, M. (1978). Thermal Interactions among the Confining Walls of a Turbulent Recirculating Flow, *Int. J. Heat Mass Transfer*, vol. 21, p. 269.

Peaceman, D. W. and Rachford, H. H. (1955). The Numerical Solution of Parabolic and Elliptic Differential Equations, *J. Soc. Ind. Appl. Math.*, vol. 3, p. 28.

Pearson, C. E. (1965). A Computational Method for Viscous Flow Problems, *J. Fluid Mech.*, vol. 21, p. 611.

Potter, D. E. and Tuttle, G. H. (1973). The Construction of Discrete Orthogonal Coordinates, *J. Comp. Phys.*, vol. 13, p. 483.

Prandtl, L. (1945). Über ein neues Formelsystem für die ausgebildete Turbulenz, *Nach. Akad. Wiss. Göttingen Math. Phys.*, p. 6.

Pratap, V. S. and Spalding, D. B. (1975). Numerical Computation of the Flow in Curved Ducts, *Aeronaut. Q.*, vol. 26, p. 219.

Pratap, V. S. and Spalding, D. B. (1976). Fluid Flow and Heat Transfer in Three-Dimensional Duct Flows, *Int. J. Heat Mass Transfer*, vol. 19, p. 1183.

Raithby, G. D. (1976a). A Critical Evaluation of Upstream Differencing Applied to Problems Involving Fluid Flow, *Comp. Methods Appl. Mech. Eng.*, vol. 9, p. 75.

Raithby, G. D. (1976b). Skew Upstream Differencing Schemes for Problems Involving Fluid Flow, *Comp. Methods Appl. Mech. Eng.*, vol. 9, p. 153.

Raithby, G. D. and Torrance, K. E. (1974). Upstream-Weighted Differencing Schemes and Their Application to Elliptic Problems Involving Fluid Flow, *Comput. Fluids*, vol. 2, p. 191.

Ramsey, J. W. and Goldstein, R. J. (1970). Interaction of a Heated Jet with Deflecting Stream, NASA CR-72613.

Rastogi, A. K. and Rodi, W. (1978). Calculation of General Three-Dimensional Turbulent Boundary Layers, *AIAA J.*, vol. 16, p. 151.

Runchal, A. K. (1972). Convergence and Accuracy of Three Finite Difference Schemes for a Two-Dimensional Conduction and Convection Problem, *Int. J. Num. Methods Eng.*, vol. 4, p. 541.

Runchal, A. K. and Wolfshtein, M. (1969). Numerical Integration Procedure for the Steady State Navier-Stokes Equations, *J. Mech. Eng. Sci.*, vol. 11, p. 445.

Scarborough, J. B. (1958). *Numerical Mathematical Analysis*, 4th ed., Johns Hopkins Press, Baltimore.

Spalding, D. B. (1972). A Novel Finite-Difference Formulation for Differential Expressions Involving Both First and Second Derivatives, *Int. J. Num. Methods Eng.*, vol. 4, p. 551.

Spalding, D. B. (1977). *GENMIX—A General Computer Program for Two-Dimensional Parabolic Phenomena*, Pergamon, New York.

Sparrow, E. M., Baliga, B. R., and Patankar, S. V. (1978). Forced Convection Heat Transfer from a Shrouded Fin Array with and without Tip Clearance, *J. Heat Transfer*, vol. 100, p. 572.

Sparrow, E. M. and Patankar, S. V. (1977). Relationships among Boundary Conditions and Nusselt Numbers for Thermally Developed Duct Flows, *J. Heat Transfer,* vol. 99, p. 483.

Sparrow, E. M., Patankar, S. V., and Ramadhyani, S. (1977). Analysis of Melting in the Presence of Natural Convection in the Melt Region, *J. Heat Transfer,* vol. 99, p. 520.

Sparrow, E. M., Patankar, S. V., and Shahrestani, H. (1978). Laminar Heat Transfer in a Pipe Subjected to a Circumferentially Varying External Heat Transfer Coefficient, *Num. Heat Transfer,* vol. 1, p. 117.

Stone, H. L. (1968). Iterative Solution of Implicit Approximations of Multi-Dimensional Partial Differential Equations, *SIAM J. Num. Anal.,* vol. 5, p. 530.

Winslow, A. M. (1966). Numerical Solution of the Quasilinear Poisson Equation in a Nonuniform Triangle Mesh, *J. Comp. Phys.,* vol. 1, p. 149.

INDEX